Emerging Internet-Based Technologies

Emerging Internet-Based Technologies

Matthew N. O. Sadiku

CRC Press
Taylor & Francis Group
Boca Raton London New York

CRC Press is an imprint of the
Taylor & Francis Group, an **informa** business

CRC Press
Taylor & Francis Group
6000 Broken Sound Parkway NW, Suite 300
Boca Raton, FL 33487-2742

International Standard Book Number-13: 978-0-367-03029-2 (Hardback)

Visit the Taylor & Francis Web site at
http://www.taylorandfrancis.com

and the CRC Press Web site at
http://www.crcpress.com

To my brother and his wife:

Moses and Victoria Sadiku Ojogiri

Contents

Preface

The industrial revolution changed the dynamics of our society through rapid urbanization and rise of cities, working women, rise of the middle class, and creation of job opportunities. The Internet revolution predominately brought with it changes that were not only technological but societal and pervasive in scope. The Internet today is a widespread information infrastructure, which is often called the Information Superhighway. It is regarded by many as the greatest technological disruption of all time.

The author of this book has identified the seven key emerging Internet-related technologies: Internet of things, smart everything, big data, cloud computing, cybersecurity, software-defined networking, and online education. Taken together, these technologies are transformational and disruptive. This book provides researchers, students, and professionals a comprehensive introduction, applications, benefits, and challenges for each technology. It presents the impact of these cutting-edge technologies on our global economy and its future. The word "technology" refers to a "collection of techniques, skills, methods, and processes used in the production of goods or services."[1]

The author was motivated to write this book partly due to the lack of a single source of reference on these new technologies. These are monographs on each technology, but there is none that combines the technologies. Hence, the book will help provide a beginner with an introductory knowledge about these emerging technologies. The author's main objective is to provide a concise treatment that is easily digestible. Since each topic would require a whole book by itself, the book only focuses on core ideas and basic concepts. It is hoped that the book will be useful to practicing engineers, computer scientists, and information business managers.

The book has seven chapters. Chapter 1 deals with the Internet of things (IoT), which is a worldwide network of physical objects using the Internet as a communication network. The Internet has changed everything and provided "smartness" to these connected things. IoT is the Internet of the future and it will seriously impact our life.

Chapter 2 is on "smart everything," which is enabled by the deployment of IoT. Smart technologies produce smart cities, smart homes, smart energy, smart transportation, smart manufacturing, smart agriculture, smart living, smart environment, and smart university. These technologies will ensure equity, fairness, and realize a better quality of life.

Chapter 3 is devoted to "big data," which is an emerging term used in business, engineering, and other domains. Big data is a high-volume, high-velocity, and high-variety information that requires special information processing tools. The ability to collect and analyze huge amounts of data is a growing problem in every domain. Big data is poised to have an increasing impact in all sectors of our society.

In Chapter 4, we cover cloud computing, which is an emerging computing paradigm for delivering computing services (such as servers, storage, databases, networking, software, analytics, and more) over the "the cloud" or Internet with pay-as-you-go pricing. It is a disruptive, global technology that is massively changing how computing is done.

Chapter 5 introduces cybersecurity, which is the act of protecting information systems from all sources of attack: criminal syndicates, cyber vandals, terrorists, identity thieves, intruders, and disgruntled insiders/employees. Since weak cybersecurity endangers our country, cybersecurity has become a national imperative and a government priority.

Chapter 6 introduces software-defined networking, which is the next wave of networking. It addresses the failure of the IP traditional networks to support the dynamic, scalable computing and storage needs of today's applications. It is becoming increasing important for the next generation networks such as Internet backbone and data center networks.

Finally, Chapter 7 addresses online education, which is currently the latest, most popular form of distance education. The greatest appeal for online education is its convenience, accessibility, low cost, and availability to learners. With the advances in Internet technologies, online education has gained a lot of popularity in recent times and plays a formidable role in US higher education's goal of meeting the demands of the 21st century. It is the future of education.

I appreciate the support received from Dr. Kelvin Kirby, head of department of electrical and computer engineering at Prairie View A&M University.

I would like to thank Nora Konopka, Emeline Jarvie, and other staff of CRC for their help. I want to thank my wife for her understanding, support, and prayer.

Matthew N. O. Sadiku

Reference

1. https://en.wikipedia.org/wiki/Technology

Acronyms

AIOTI	Alliance for Internet of Things Innovation
AWS	Amazon Web Services
BAN	Body area network
BD	Big data
BDA	Big data analytics
BLE	Bluetooth Low Energy
BYOD	Bring your own device
CC	Cloud computing
CFAA	Computer Fraud and Abuse Act
CMS	Course management system
CoT	Cloud of Things
DHS	Department of Homeland Security
DM	Data mining
DL	Deep learning
EEA	Economic Espionage Act
ERC	European Research Cluster
FAO	Food and Agriculture Organization
FCC	Federal Communications Commission
GIS	Geographic information system
GPS	Global positioning system
HAN	Home area network
IaaS	Infrastructure-as-a-Service
ICT	Information and communication technology
IDS	Intrusion detection system
IEEE	Institute of Electrical and Electronics Engineers
IEFT	Internet Engineering Task Force
IIoT	Industrial Internet of Things
IoT	Internet of Things
IoV	Internet of Vehicles
IPS	Intrusion prevention system
ITS	Intelligent transport system
LAN	Local area network
MEMS	Micro-Electro-Mechanical Systems
ML	Machine learning
MOOC	Massive Online Open Course
NAN	Neighborhood area network
NFC	Near-field communication
NOS	Network operating system
ONF	Open Networking Foundation
OS	Operating system

PaaS	Platform-as-a-Service
PAN	Personal area network
RFID	Radio-frequency identification
SaaS	Software-as-a-Service
SC	Smart city
SCA	Smart-climate agriculture
SCADA	Supervisory control and data acquisition
SDN	Software-defined networking
SIoT	Social Internet of Things
SM	Smart manufacturing
STA	Smart Transportation Alliance
TCP/IP	Transmission Control Protocol/Internet Protocol
UAV	Unmanned aerial vehicle
UN	United Nations
VANET	Vehicular ad hoc network
WAN	Wide area network
WSN	Wireless sensor network

Author

Matthew N. O. Sadiku received his B.Sc. degree in 1978 from Ahmadu Bello University, Zaria, Nigeria and his M.Sc. and Ph.D. degrees from Tennessee Technological University, Cookeville, TN in 1982 and 1984, respectively. From 1984 to 1988, he was an assistant professor at Florida Atlantic University, Boca Raton, FL, where he graduated with a degree in computer science. From 1988 to 2000, he was at Temple University, Philadelphia, PA, where he became a full professor. From 2000 to 2002, he was with Lucent/Avaya, Holmdel, NJ as a system engineer and with Boeing Satellite Systems, Los Angeles, CA as a senior scientist. Currently, he is a professor of electrical and computer engineering at Prairie View A&M University, Prairie View, TX.

He is the author of over 510 professional papers and 80 books including *Elements of Electromagnetics* (Oxford University Press, 7th ed., 2018), *Fundamentals of Electric Circuits* (McGraw-Hill, 6th ed., 2017, with C. Alexander), *Computational Electromagnetics with MATLAB* (CRC, 4th ed., 2018), *Metropolitan Area Networks* (CRC Press, 1995), and *Principles of Modern Communication Systems* (Cambridge University Press, 2017, with S. O. Agbo). In addition to the engineering books, he has written Christian books including *Secrets of Successful Marriages, How to Discover God's Will for YourLife*, and commentaries on all the books of the New Testament Bible. Some of his books have been translated into French, Korean, Chinese (and Chinese Long Form in Taiwan), Italian, Portuguese, and Spanish.

He was the recipient of the 2000 McGraw-Hill/Jacob Millman Award for outstanding contributions in the field of electrical engineering. He was also the recipient of Regents Professor award for 2012–2013 by the Texas A&M University System. He is a registered professional engineer and a fellow of the Institute of Electrical and Electronics Engineers (IEEE) "for contributions to computational electromagnetics and engineering education." He was the IEEE Region 2 Student Activities Committee Chairman. He was an associate editor for IEEE Transactions on Education. He is also a member of Association for Computing Machinery (ACM) and American Society of Engineering Education (ASEE). His current research interests are in the areas of computational electromagnetic, computer networks, and engineering education. His works can be found in his autobiography, "My Life and Work" (Trafford Publishing, 2019), or on his website: www.matthewsadiku. com. He currently resides with his wife Kikelomo in Hockley, TX. He can be reached via email at sadiku@ieee.org.

1

Internet of Things

The best way to predict the future is to invent it.

Alan Kay

1.1 Introduction

The Industrial Revolution changed the dynamics of a society through rapid urbanization and rise of cities, working women, rise of the middle class, and creation of job opportunities. The Internet Revolution predominately brought with it changes that were not only technological but societal and pervasive in scope. The Internet is regarded by many as the greatest technological disruption of all time. Access to information combined with global ecommerce reshaped established conventions. The next revolution that will disrupt our lifestyle, technology, and business is the Internet of Things (IoT) [1].

Most of the devices we use today support communication technology. Such devices include cell phones, sensors, smart grid, and laptops. These devices can interact among themselves through the Internet. Such a paradigm is called the "Internet of Things" or "Internet of Objects," where the devices are referred to as the things. It is the Internet of relating to things. A thing in the IoT environment is always connected with the external and interacting with other things, such as objects, humans, animals, or plants.

The IoT is a link between objects in the real world with the virtual world, thereby enabling anytime, anywhere connectivity for anything. The goal of IoT is to integrate and automate everything from home appliances to plants on factory floors. Experts predict that the IoT will consist of about 50 billion objects by 2020. Thus, the number of things connected to the Internet now exceeds the number of people on the planet. IoT is the next step in the evolution of the Internet, since it takes into consideration all devices connected to it. It allows all types of elements (sensors, actuators, personal electronic devices, laptops, tablets, digital cameras, smart phones, alarm systems, home appliances, industrial machines, etc.) to autonomously interact with each other. Today, big players from a wide range of industries are adopting IoT and basing their future business and growth prospects on it.

This chapter begins by presenting the basic characteristics of IoT. Then it presents the technologies enabling IoT and some popular applications of IoT. It gives the current trends in IoT: industrial IoT, Internet of Vehicles (IoV), and social IoT. It addresses some challenges facing IoT and some organizations working on some of these challenges. Finally, the last section provides the conclusions.

1.2 IoT Basics

The term, IoT, was first coined by Kevin Ashton, a British entrepreneur in 1999. He meant to represent the concept of computers and machines with sensors, which are connected to the Internet to report status and accept control commands [2].

IoT (also known as sensor network or Industrial Internet) is a global network infrastructure of interconnected devices (such as sensors, actuators, personal electronic devices, laptops, tablets, digital cameras, smart phones, alarm systems, home appliances, or industrial machines, and other smart devices) that are enabled with technology of interacting and communicating with each other. It mainly enables the interconnection of Thing to Thing (T2T), Human to Thing (H2T), and Human to Human (H2H). By collecting and combining data from various IoT devices and using big data analytics, decision-makers can take appropriate actions with important economic, social, and environmental implications.

As shown in Figure 1.1 [3], the IoT can be divided into three layers: perception (or sensing) layer, network layer and application layer:

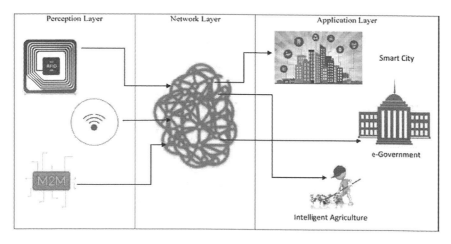

FIGURE 1.1
The IoT layers [3].

1. The *perception layer* collects from devices such radio-frequency identification (RFID) tags and readers, camera, GPS, and sensors. In this layer, the wireless smart systems with sensors can automatically sense and exchange information among different devices and remotely control them.

2. The *network layer* is mainly messaging and processing information. The role of this layer is to connect all things together and allow them to share the information with each other.

3. The *application layer* is the IoT and the application systems.

The fundamental characteristics of IoT include interconnectivity, things-related services, scalability, heterogeneity, and dynamic changes.

1.3 Enabling Technologies

Translating the IoT concept into the real world requires integration of several enabling technologies. Technologies that enable IoT are shown in Figure 1.2 [4]. These IoT technologies are widely used for the deployment of successful IoT-based products and services. These include RFID, wireless sensor networks (WSN), middleware, cloud computing, artificial intelligence, IoT application software, and Micro-Electro-Mechanical Systems (MEMS) [5–7].

1. *Networking technologies.* These can be wired or wireless. The choice depends on the geographical range to be covered. Ethernet and fiber optics are examples of wired technologies, while Wi-Fi is a typical wireless technology. Bluetooth was introduced in 1999 as a wireless technology for transferring data over short distances. Newly released smart phones have Bluetooth Low Energy (BLE) hardware

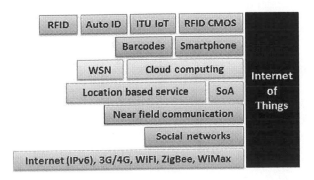

FIGURE 1.2
Technologies associated with IoT [4].

in them. ZigBee is a low-power LAN protocol based on the IEEE 802.15.4 standard. It is specifically designed for low-rate sensors.

2. *WSN*. This is perhaps the most important integral part of IoT. Sensors are usually connected as WSNs to monitor physical properties in specific environments. The purpose of a sensor is to convert a non-electric signal to an electrical signal that can be transmitted through an electric circuit. An actuator complements a sensor; it converts an electrical signal into physical action. IoT may be regarded as a things-connected network, where things are wirelessly connected via smart sensors. These sensors collaborate with each other to provide information of their physical surroundings. They collect data that may enable the business make a good decision. TinyOS serves as operating systems for sensor networks due to limited resources, flexibility, and low power.

3. *RFID*. RFID is an emerging noncontact identification technology which is designed to enhance the bar code technology. It refers to technologies that use radio waves to identify objects, animals, or people. It enables the design of microchips for wireless communication that can identify anything they are attached to automatically just as an electronic barcode. It is now widely used for tracking objects, humans, or animals. A typical RFID system consists of three main components: tags, readers, and controller [8]. Optical tags are used for low-cost tagging.

4. *Near-field communication* (NFC). NFC is a subset within the family of RFID. As the name implies, NFC is a set of short-range wireless technologies that enable two electronic devices to communicate full-duplex and typically require a separation of 10 cm or less between them. NFC employs electromagnetic induction between two antennas and operates within the globally available unlicensed radio-frequency ISM band of 13.56 MHz. It allows a device, usually a mobile phone, to collect data from another device at close range without the need for an Internet connection. Virtually every mobile operating system maker has their own apps that offer unique NFC functionality.

5. *MEMS*. This field, which encompasses all aspects of science and technology, is involved with things on a smaller scale. MEMS technology gives scientists and engineers the tools to build things that have been impossible or prohibitively expensive with other technologies. It is a manufacturing technology; a new way of making complex electromechanical systems using batch fabrication techniques similar to the way integrated circuits (ICs) are made and making these electromechanical elements along with electronics [9]. This is an enabling technology for IoT because MEMS manufacturing produces small, low-cost sensors and actuators. MEMS technology has

already taken root firmly in today's world. It is destined to become a hallmark 21st century manufacturing technology with numerous and diverse applications.

These technologies have helped capture of and access to real-time information. IoT is the convergence of the Internet with RFID, sensors, and smart objects. IoT has potential for societal as well as economic impact. Potential applications are the major force driving the future of the IoT. Some see IoT as a vehicle for economic growth.

1.4 Applications

The main advantage of the IoT concept is the high impact it will have on various aspects of everyday-life of potential users. The US government and government of other nations such as UK, China, and India are taking drastic measures to incorporate IoT into many industries. This action will have impact on various government projects such as smart cities, smart manufacturing (SM), smart materials, smart transportation, smart power grids, smart agriculture, etc. The IoT is applied in health care, automotive industries, smart power grid, manufacturing, transportation, agriculture, logistics, pharmaceutics, surveillance, etc. Some of these applications are illustrated in Figure 1.3 [10]. Here we will consider some typical applications of IoT [11,12].

FIGURE 1.3
Typical IoT applications [10].

- *Manufacturing.* This is the largest industry from an IoT spending (software, hardware, connectivity, and services) perspective. Manufacturing is among the industrial sectors that will be directly impacted by the disruption springing from industrial IoT. By collecting information from IoT sensors and analyzing it, SM increases production efficiency in factories. IoT is posed to automate, monitor, and analyze information from machinery and equipment in exciting new business ventures. Thus, IoT provides numerous great opportunities to advance manufacturing in achieving better system performances in globalized and distributed environments.

- *Transportation.* This represents the second largest IoT market. Today's transportation infrastructure is stressed to the breaking point. Many cities have begun smart transportation initiatives to optimize their public transportation routes, create safer roads, reduce infrastructure costs, and alleviate traffic congestion. Airlines, rail companies, and public transit agencies can aggregate huge quantities of data to optimize operations.

- *Energy management.* Energy costs account for major production and monitoring. IoT can be put to use in this regard. Sensors connected to IoT can be used to monitor energy consumption, help control costs, and ensure compliance. IoT simplifies the process of energy monitoring and management while maintain low cost and high precision.

- *Cybersecurity.* Since industrial plants may be potential targets for terrorist attacks, cybersecurity should be a top priority. Companies must consider physical security, industrial systems security, and data security. Using IoT, companies can bring data together easily, enabling them to quickly identify unauthorized access and prevent further intrusion. This way they can prevent hackers from gaining access to their sensitive data.

- *Technologies.* The IoT provides big opportunities for using technologies at affordable cost. It is helping to simplify business processes and provide more insight that can help improve quality, reduce downtime, decrease the costs of maintenance, and increase on-time delivery. Using the technologies to connect devices, plants, assets, people, products, and processes is helping industries to make data-driven decisions and stay competitive.

- *Healthcare.* IoT technology can enhance existing healthcare systems and the general practice of medicine. It exploits integrated network of sophisticated medical devices and drastically improve medical research and emergence care. It empowers healthcare professionals to use their knowledge and experience to solve problems. It also promises to benefit disabilities and the elderly at a reasonable cost.

- *Governments.* Governments are already using IoT to develop smart homes, smart cities, and smart nations. They also use it for remote monitoring and control of devices within the city boundary. Governments can use IoT to improve law enforcement, planning, defense, and economic management. IoT supports cities through major services and infrastructure such as healthcare and transportation.

Other applications include digital supply chain, education, retail, automotive, unmanned aerial vehicles or drones, aerospace, agriculture, predictive maintenance, data-enabled services, connected logistics, quality assurance, integrated supply chain, logistics, and smart environment, smart homes, smart grid, smart city, smart farming, defense, entertainment, industry 4.0, processing industries such as oil and gas, energy consumption optimization, safety, and health monitoring of workers.

1.5 Current Trends in IoT

The growth of the IoT is drastically making impact on home and industry. New areas of IoT include industrial IoT, IoV, and social IoT. These may be regarded as extended applications of IoT.

1.5.1 Industrial Internet of Things

While the IoT affects transportation, healthcare, or smart homes, the Industrial Internet of Things (IIoT) refers in particular to industrial environments. IIoT is a new industrial ecosystem that combines intelligent and autonomous machines, advanced predictive analytics, and machine–human collaboration to improve productivity, efficiency, and reliability. It is bringing about a world where smart, connected embedded systems and products operate as part of larger systems.

The term "industrial Internet" is strongly pushed by General Electric. Some see this as the biggest and most important part of the overall IoT picture. In fact, there are two subsets of IoT: the Consumer IoT and the Industrial IoT. The Consumer IoT naturally evolves from human-operated computers to automated things that surround humans. It consists of smart home devices, wearable computers, cameras, and networked appliances. The IIoT refers to a large number of interconnected industrial systems that are communicating, sharing data, and improving industrial performance to benefit the society. It includes networked smart power grid, manufacturing, medical and transportation infrastructures. It requires high reliability, lower power usage, and timely exchange of information [13]. It is helping to improve productivity,

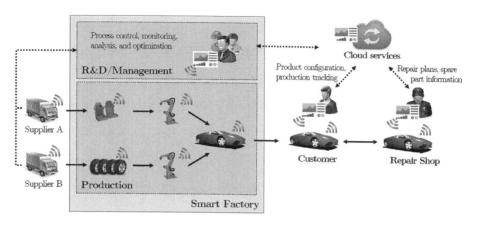

FIGURE 1.4
A typical industrial Internet of Things [14].

enhance worker safety, and reduce operating costs. A typical IIoT is shown in Figure 1.4 [14].

The IIoT refers to the application of IoT across several industries such as manufacturing, logistics, oil and gas, transportation, energy/utilities, chemical, aviation, and other industrial sectors. This killer application is poised to revolutionize the world. Companies that want to stay competitive should embrace the IIoT as soon as possible.

1.5.2 Internet of Vehicles

As more and more people drive cars and vehicles, there is a corresponding increase in the number of fatalities that occur due to accidents. As these vehicles are increasingly being connected to the IoT, they form the IoV. Thus, IoV is the convergence of the mobile Internet and IoT. It is an emerging field for the automotive industry and an important part of the smart cities. The new area of IoT is driving the evolution of conventional vehicle ad-hoc networks in IoV. IoV technology refers to dynamic mobile communication systems that communicate between vehicles and public networks using V2V (vehicle-to-vehicle), and (vehicle-to-sensor) interactions. Research on IoV is carried out by different enterprises and organizations. Challenges facing IoV and slowing down its adoption include big data, security, privacy, reliability, mobility, and standards [15].

For online presence of vehicles, each vehicle should have a uniquely identifiable number of the Internet. A vehicular global ID (GID) terminal is at the core of the IoV. Simply put, the GID addresses problems with RFID that include its one-way nature, limited range and coverage, lack of speed, passive and unintelligent operation.

IoV has become an indispensable platform with information interaction among vehicles, humans, and road-side infrastructures. A vehicle will be a sensor platform, absorbing information from the environment, from other vehicles, from the driver and using it for safe navigation, pollution control, and traffic management.

The notion of IoV is no longer a matter of IT applications in the automotive industry; it has become a national and global concern. With time IoV will become important part of us and make intelligent transportation systems do without traffic lights, road accidents, and other related problems. It will make millions of people enjoy more convenient, comfortable, and safe traffic service [16].

1.5.3 Social Internet of Things

The Social Internet of Things (SIoT) is an offshoot of the IoT, where social relationships are established among objects, things, and people. In other words, SIoT is an IoT where things are capable of establishing social relationships with other objects autonomously. It integrates social networking concepts into the IoT.

The main objective of IoT is to equip objects with computing and communication power, so that they can interact with each other. A typical system architecture of the SIoT is shown in Figure 1.5 [17]. The main characteristic of SIoT is that the objects are capable of establishing social relationships with each other. Social relationships can be established between people and things, and between things and things based on their experiences, preferences, and requirements without underlying network protocols. Trust and privacy are the two factors that govern these relationships and have a huge impact on SIoT.

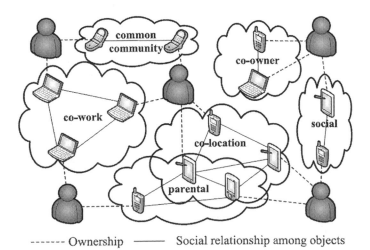

FIGURE 1.5
Architecture of the social Internet of Things [17].

Applications of SIoT include smart cities, military, marketing, health, and IoV. Without doubt, interconnected intelligent devices and SIoT will influence almost all aspects of our lives.

1.6 Benefits of IoT

In a connected world, IoT has enabled sophisticated monitoring, control, and communication. For IoT, its real-world benefits are beginning to shine through. It presents several benefits to consumers and has the potential to change the ways they interact with Internet. First, the IoT is not only a new way of gathering data but also a way to gather new facts. Second, IoT technologies tend to be open, flexible, and easy to build upon. Third, the falling cost of sensors and RFID and the availability of wireless networks (such as Wi-Fi) have opened up new opportunities for IoT technology. Fourth, governments are saving costs with IoT by creating smart cities and smart transportation.

Those who implement IoT in the right way have found their expectations surpassed.

The drawbacks/challenges of IoT are mentioned in the next section.

1.7 Challenges

IoT is the biggest promise of the technology today, but a number of challenges need to be overcome in order for IoT to achieve its objective. Some scholars and social observers doubt whether the promises of the ubiquitous computing revolutions would come true. Numerous challenges stand in the way of the vision becoming a reality: security, privacy, standardization, interoperability, cost, addressing mechanism, and data management. We will consider six major challenges IoT faces [18]:

- *Security*. A major challenge facing IoT is cybersecurity and data security, which is rising in importance due to increased vulnerability to attacks and data breaches. Anything connected to the Internet can be hacked. Security issues such as access control, secure communication, and secure data storage in IoT environment are becoming challenging [19]. In the IoT context, data are considered sensitive because data will encapsulate various aspects of industrial operation, including highly sensitive information about products, business strategies, and companies. Sensitive data may warrant an extended use of privacy torts. IoT presents special security risk because it depends on both the public Internet and private intranets. Each layer of the IoT is vulnerable to attacks. People's concern is that the IoT is being developed rapidly without due consideration of the profound security challenges involved. Ensuring security of IoT products and services must be given a priority and this requires collaboration across borders, sectors, and organizations. Government can serve as catalysts for developing good IoT security practices.

- *Privacy.* The IoT can challenge the traditional expectations of privacy. The user may not be aware that an IoT device is collecting and sharing data about the user with third parties. In a world where all things are connected, individual's right to privacy needs to be protected. In order to achieve a reliable and secure IoT environment, a number of privacy principles and security protocols must be implemented. Principles of informed consent, data confidentiality and security must be safeguarded. Ensuring privacy rights and respect for users is needed for trust and confidence in IoT services. Privacy is also a major concern in important enablers of the IoT technology such as RFID, WSNs, and mobile applications [20].

- *Addressing mechanism.* Integration of every device with the Internet necessitates that devices use an IP address as a unique identifier. To some extent, the future of IoT will be limited without the support of IPv6. The supply of IPv4 addresses was exhausted in February 2014. The deployment of IPv6 will cover this IP address shortage by assigning addresses to billions of devices and make network management less complex. IPv6 will make network management easier due to auto-configuration capabilities and improve security features.

- *Cost.* Cost of deployment, operation, and maintenance of the network is an important factor. Recently, RFID tags and encryption technologies have become available at reasonable prices.

- *Energy consumption.* Energy consumption is a major concern because devices connected to IoT consume power. Providing power to sensors for a prolonged period of time is key to IoT being deployed successfully. Sensors must be self-sustaining because it is impossible to change batteries in billions of IoT devices around the globe. Transfer, storage, and processing of data are the major energy-consuming activities with an IoT. Improving energy efficiency has positive economic and environmental impacts and reduces the operational cost.

- *Standardization/interoperability.* Standardization and interoperability between heterogeneous, complex Internet-connected objects are always a challenge. Standards are vital to ensure that any new device added to the infrastructure can interact with existing equipment. Interoperability is the cornerstone of the open Internet. The IoT is a very complex heterogeneous network because it involves various types of networks with various communication technologies. This leads to a related problem of lack of common platform or technological standards in some areas, while other sectors are divided and fragmented. Standardization of technologies involved is important for system integration. This involves standardizing RFID technologies, sensor networks, NFC technology, MEMS, communication protocols, and IoT. IEEE and the Internet of Things Global Standards Initiative are attempting to develop IoT standards. Standards on IoT have attracted a lot of attention in many countries. Global standards are needed to achieve economy of scale and global networking. They are the key to interoperability.

- *Data management.* This is a challenge because IoT sensors and devices are generating massive amounts of data that need to be processed and stored. Data collected from IoT devices may be transmitted from one jurisdiction to another with no roadblocks. Information may be transmitted across borders without the user knowing it. The application of IoT devices raises legal and regulatory issues that did not exist prior to these devices.

Despite the challenges, the capabilities of the IoT are already quite impressive.

1.8 Organizations Working on IoT

A wide range of organizations, alliances, and government efforts around the globe are addressing issues rated to IoT and some are providing standardization. These include the following [21]:

1. *AIOTI.* The Alliance for Internet of Things Innovation was launched by the European Commission and various key IoT players in 2015 to support the development of IoT ecosystem.

2. *AllSeen Alliance.* This group promotes widespread adoption and application of IoT. It provides an open source IoT framework for the Internet of Everything with contributions from member companies and the open source community.

3. *IEEE.* The Institute of Electrical and Electronics Engineers (IEEE) has launched the IoT Technical Community, with membership comprised of those involved in research, implementation, application and usage of this Internet-enabled vision of our future. It also devotes a journal to IoT: Internet of Things Journal.

4. *ERC.* The European Research Cluster coordinates activities on IoT across Europe. Its objectives include facilitating networking of

IoT-related projects and research in Europe and leveraging expertise, talents, and resources for maximum impact.

5. *Internet Engineering Task Force* (IEFT). This standards-setting body has
   an IoT Directorate monitoring IoT-related activities in other standard
   groups. Different IEFT groups are working on standardization [22].

6. *IIC.* The Industrial Internet Consortium was formed in 2014 to bring
   together organizations and researchers to accelerate the delivery of
   an industrial IoT. Membership includes technology innovators, market leaders, researchers, universities, and government organizations.

7. *Internet of Things Consortium.* This industry group generates opportunities to meet and collaborate with other leading companies,
   thinkers, and movers in the IoT world. It provides education of IoT
   products and services to consumers, sales channels, and investors.

In addition to these organizations, IoT is being popularized as sound business concept by industry leaders like Intel, IBM, Apple, Microsoft, Cisco,
Google, Samsung, Oracle, and Amazon.

1.9 Conclusion

The Internet has changed everything and provided "smartness" to these
connected things. Its principal goal, back in 1973, was to enable computers
to communicate with each other. Its usage is multifaceted and is expand-
ing on a daily basis. The IoT is emerging as a disruptive technology of the
21st century—an extension of the Internet Revolution, which has made a tremendous impact on social and business environments. The goal of IoT is to
enable things to be connected anytime, anyplace, with anything and anyone
using any network and any service. Its development is gaining more and
more momentum.

The IoT is a worldwide network of physical objects using the Internet as
a communication network. It is the next stage of information revolution
because it deals with the interconnectivity of everything. It is becoming what
people on the street can relate to. Through several technological advances,
the modern society is heading toward an "always connected" paradigm.

IoT is interdisciplinary in nature because it embraces many technologies,
such as computer networks, sensor technology, embedded technology, and
wireless communication technology. The multifaceted nature of IoT makes
its hard to develop a curriculum on IoT. Consequently, current academic programs do not meet the rapid demand for IoT skilled professionals.

IoT is the Internet of the future. It will seriously impact our life. Connected
smart devices will do more, do it smarter, and do it faster [23]. More

information about IoT can be found in some recent books [24–26], several books on it available on Amazon.com, and *IEEE Internet of Things Journal*, which is exclusively devoted to IoT.

References

1. R. Bhatnagar, "Role of IoT & its impact on various industries in India," *DataQuest*, April 2016.
2. M. N. O. Sadiku, S. M. Musa, and S. R. Nelatury, "Internet of Things: an introduction," *International Journal of Engineering Research and Advanced Technology*, vol. 2, no. 3, March 2016, pp. 39–43.
3. H. Li and Y. Liu, "Research on the chemical logistics management information platform based on Internet of Things," *Proceedings of International Conference on E-Product E-Service and E-Entertainment*, November 2010.
4. S. S. Tabrizi and D. Ibrahim, "Security of the Internet of Things: an overview," *Proceedings of the 2016 International Conference on Communication and Information Systems*, December 2016, pp. 146–150.
5. "Internet of Things," *Wikipedia*, https://en.wikipedia.org/wiki/Internet_of_Things.
6. S. Greengard, *The Internet of Things*. Cambridge, MA: The MIT Press, 2015.
7. S. Li, L. D. Xu, and S. Zhao, "The Internet of Things: a survey," *Information System Front*, vol. 17, no. 2, 2015, pp. 243–259.
8. M. N. O. Sadiku, A. E. Shadare, and S. M. Musa, "A primer on RFID," *International Journal of Recent Advances in Multidisciplinary Research*, vol. 2, no. 11, December 2016, pp. 17–20.
9. M. N. O. Sadiku, "MEMS: A breakthrough technology," *IEEE Potentials*, vol. 21, no. 1, February/March 2002, pp. 4–5.
10. K. K. Patel and S. M. Patel, "Internet of Things – IoT: definition, characteristics, architecture, enabling technologies, application & future challenges," *International Journal of Engineering Science and Computing*, vol. 6, no. 5, May 2016, pp. 6122–6131.
11. "Internet of Things tutorial," www.tutorialspoint.com//internet_of_things/index.htm.
12. M. N. O. Sadiku, S. M. Musa, and O. S. Musa, "Internet of Things in the chemical industry," *International Journal of Advances in Scientific Research and Engineering*, vol. 3, no. 10, November 2017, pp. 24–28.
13. M. N. O. Sadiku, Y. Wang, S. Cui, and S. M. Musa, "Industrial Internet of Things," *International Journal of Advances in Scientific Research and Engineering*, vol. 3, no. 11, December 2017, pp. 1–4.
14. A. R. Sadeghi1, C. Wachsmann, and M. Waidner, "Security and privacy challenges in industrial Internet of Things," *Proceedings of the 52nd Annual Design Automation Conference*, June 2015.
15. F. Wang et al., "Architecture and key technologies for Internet of vehicles: a survey," *Journal of Communications and Information Networks*, vol. 2, no. 2, June 2017, pp. 1–17.

16. M. N. O. Sadiku, M. Tembely, and S. M. Musa, "Internet of vehicles: an introduction," *International Journal of Advanced Research in Computer Science and Software Engineering*, vol. 8, no. 1, January 2018, pp. 11–13.

17. Z. Chen et al., "A scheme of access service recommendation for the Social Internet of Things," *International Journal of Communication Systems*, vol. 29, 2016, pp. 694–706.

18. I. F. Akyildiz et al., "The internet of bio-nano things," *IEEE Communications Magazine*, vol. 53, no. 3, March 2015, pp. 32–40.

19. M. Conti et al., "Internet of Things security and forensics: challenges and opportunities," *Future Generation Computer Systems*, vol. 78, 2018, pp. 544–546.

20. J. H. Ziegeldort, O. G. Morchon, and K. Wehrle, "Privacy in the Internet of Things: threats and challenges," *Security and Communication Networks*, vol. 7, 2014, pp. 2728–2742.

21. K. Rose, S. Eldrigge, and L. Chapin, "The Internet of Things: an overview: understanding the issues and challenges of a more corrected world," The Internet Society, October 2015, www.internetsociety.org/wp-content/uploads/2017/08/ISOC-IoT-Overview-20151221-en.pdf.

22. I. Ishaq et al., "IETF standardization in the field of the Internet of Things (IoT): a survey," *Journal of Sensor and Actuator Networks*, vol. 2, 2013, pp. 235–287.

23. M. Miller, *The Internet of Things: How Smart TVs, Smart Cars, Smart Homes, and Smart Cities are Changing the World*. Indianapolis, IN: Que, 2015.

24. H. Geng (ed.), *Internet of Things and Data Analytics Handbook*. Hoboken, NJ: John Wiley & Sons, 2017.

25. R. Buyya and A. V. Dastjerdi (eds.), *Internet of Things: Principles and Paradigms*. Cambridge, MA: Morgan Kaufmann, 2016.

26. D. Norris, *The Internet of Things: Do-It-Yourself Projects with Arduino, Raspberry Pi, and BeagleBone Black*. New York: McGraw-Hill Education, 2015.

2

Smart Everything

Two things define you. Your patience when you have nothing, and your attitude when you have everything.

Anonymous

2.1 Introduction

The Internet will serve as a catalyst for much of our innovation and prosperity in all sectors in the future. The Internet of Things (IoT) points to something even more exciting: the concept of "smart everything." The new concept takes the IoT one step further. From smartphones to smart everything, the digital technology is moving us to the threshold of *another* revolution.

Smart technologies exist at the intersection of digital technology and disruptive innovation. Digital technologies are affecting the human brainpower as the steam engine affected human muscle power. They produce smart cities (SCs), smart homes, smart energy, smart transportation, smart manufacturing, smart agriculture, smart living, smart environment, and smart medication. These technologies will ensure equity, fairness, and realize a better quality of life.

This chapter begins with SCs, which have been considered the wave of the future. Then it covers smart home, which is the main building block of SCs, and smart energy, which is a basic necessity in the modern society. It discusses smart transportation as an important component of the SC. It addresses smart agriculture for solving problems in the oldest sector of the US economy. Finally, it covers smart university which is an emerging concept to promote modernization of universities all over the world. It should be noted that IoT is one of the basic underlying technologies in each of these smart technologies.

2.2 Smart Cities

Cities play a major role in economic and social aspects of life worldwide. The majority of world's population resides in cities. By 2030, 4.7 billion people will live in cities. Cities are going through a new era of change. Cities' services and infrastructures are being stretched to their limits to support the population growth. Modern cities are monstrous communities, with millions of residents. They are the economic engines of the modern world because they generate economic opportunities. Cities bring individuals together and foster interchange of information by people of different cultures and skills. They collaborate, compete, and evolve together with other cities. As people change cities, cities change them.

The idea of "smart city" originated from IBM's "smart planet." There are several definitions of SCs. The word "smart" can be used to describe any device that can process information and can communicate with something. The term "smart cities" is a fuzzy concept and there is not a one-size-fits-all definition of the concept. SCs are also called intelligent cities, information cities, or virtual cities.

An SC integrates information and communication technology (ICT) in a secure manner so as to manage the city's assets. It is a high-tech urban area that connects people, information and technologies in order to increase life quality. It is one with free Wi-Fi in all public places. SCs are those communities that pursue sustainable economic development through investments in human and social capital and manage natural resources through participatory policies.

An SC is one that has digital technology embedded across all city functions. It monitors the conditions and integrates critical infrastructures such as bridges, tunnels, roads, subways, airports, seaports, and buildings.

2.2.1 Enabling Technologies

A typical architecture of an SC is shown in Figure 2.1 [1]. An SC relies on a collection of smart computing technologies. It has a wide range of electronic and digital technologies that enable its devices to communicate. Two closely related technologies, connectivity networks (IoT) and big data analytics (BDA), enable the transformation of traditional cities into SCs.

- *Connectivity networks.* The services required in an SC need the ability to gather data about its environment, infrastructure, and people. Therefore, it does not come as a surprise that the IoT represents the best way to make a city smart. IoT is the key enabler and drive of SCs. SCs have been equipped with heterogeneous electronic devices based on the IoT, which is a worldwide network of physical objects using the Internet as a communication network. The IoT is the network of interconnected devices (called things) including computers, smartphones, sensors, actuators, buildings, structures, vehicles, and wearable devices. It enables things or objects to communicate with each other and with users, becoming an integral part of the Internet. It provides the means to remotely control, monitor, and manage devices and provide actionable information from massive streams of data [2,3].

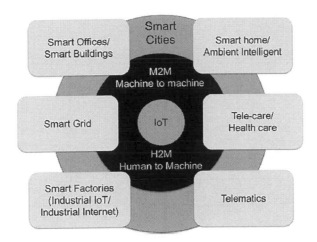

FIGURE 2.1
A typical architecture of a smart city [1].

- *Data analytics.* The large quantity of data generated by thousands of sensors and devices in an SC creates a big data (BD). The BD refers to a group of large data sets that would be hard to process using traditional data processing. It is a high-volume, high-velocity, and high-variety information that requires special information processing tools. The BDA tools are used to extract useful, actionable information from the massive streams of real-time data generated by the devices. More information on BD will be provided in Chapter 3.

2.2.2 Components of a Smart City

Components of an SC include smart people, smart governance, smart management, smart homes, smart transportation, smart infrastructure, smart buildings, smart offices, smart technology, smart energy, smart water, smart waste management, smart education, smart economy, smart mobility, smart living, smart security system, smart healthcare, smart university, smart parking, and smart environment [4]. A six-SC component system is illustrated in Figure 2.2 [5] and is explained as follows [6]:

1. *Smart government.* An SC needs a smart government to operate it. A smart government employs digital technology to make public services available online for the citizens. Government role is to integrate many technologies, stakeholders, and agency missions and bolster the concept of open government. Open government and citizen-centric governance are vital in the administration of

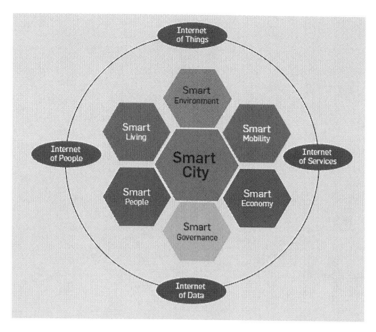

FIGURE 2.2
The main components of a smart city [5].

SCs. Open government basically includes free and open access to government information [7]. The ultimate aim of an SC is to achieve efficient, optimal management in all areas of the city—infrastructure, energy, services, education, healthcare, public safety, etc.

2. *Smart people.* While smart government provides new and improved services, citizens are the key to the development of the city. Smart people consist of social diversity, creativity, and engagement. Cities may offer online courses and services to help its citizens become smart people. Since SCs are often centers of higher education, students may play an important role. Citizens are gradually equipping their homes with IoT services such as smart TV and smart refrigerator.

3. *Smart mobility/transportation.* Transportation of people, goods, and services is of top priority for cities. Potentially, it can impact both the quality of life for citizens and the economic vitality of the cities. Transportation in big cities creates many problems such as traffic congestion, pollution, and energy consumption. Smart transportation promises to alleviate these problems. Public transport or taxi can exploit data coming from their vehicles to effectively manage their fleet. Autonomous driving technologies will save time for the user. Parking is often difficult in a large city. Smart solutions optimize the use of parking lots by equipping each parking space with a sensor which detects whether a car is parked there or not.

4. *Smart living*. This includes e-health, social services, education, surveillance systems to prevent crimes, and public safety. It involves improved standards in every aspect of living, ranging from home to workplace. The goal is to improve quality of life of the inhabitants.

5. *Smart economy*. A smart economy is a crucial to an SC since it is the main base of urban development in a smart community. This focuses on development, entrepreneurship, productivity, competitiveness, sustainability, and innovative spirit. It is expected that a wide range of smart economic service will be available for city residents. Such services include e-commerce, e-banking, utilities, and city authorities.

6. *Smart environment*. A small city is committed to its environment, which is the typical physical worlds used in human daily life. A smart environment is one that is able to autonomously acquire and apply knowledge about inhabitants and their surroundings, and adapt to improve experience. It is equipped with sensors, actuators, and computational elements. It typically includes smart homes, smart buildings, smart offices, smart hospital, etc. Cities play a major role in fighting against climate change.

Other city services such as education, healthcare, street lighting, water supply, and waste management are profoundly changing as the city becomes smart. Several models of SCs have been proposed by Cisco, IBM, Siemens, and others for different cities.

2.2.3 Benefits and Challenges

There are many benefits that result from transforming a traditional city into a smart one.

SCs act as magnets for highly educated individuals and skilled workforces. Experts claim that SCs could be efficient and more enjoyable places to live. The SC initiatives have lofty goals of improving governance and enhancing quality of life for citizens. SCs promise sustainable development, quality of services provided to residents and a high quality of life for them. They offer untold benefits for government and citizens—service provision, quality of life, and security. These benefits include integrated transport system, tourism, health, educational facilities, smart healthcare, smart energy, smart homes, crime prevention, smart infrastructure, safety and security, disaster management, and waste management.

Several initiatives all over the world have been launched to transform towns or cities from scratch to SCs. For example, India decided to set 100 SCs in May 2014. South Korea initiated 47 U-City (Ubiquitous City) projects in May 2013. So far no city has fully become an SC. Typical cities actively pursuing SC strategy include Chicago, San Francisco, San Diego, Denver,

Pittsburgh, Austin, Scottsdale, Dubuque, Ottawa, Amsterdam, Manchester, Bangalore, Lagos, Beijing, Tokyo, Montreal, and Vancouver. SCs are being pushed by big high-tech companies. Local governments now face the need to transform themselves into SCs. They must select the transformation strategy that helps them realize their ambition. Making cities smarter will continue to be important.

Building SCs has its potential barriers and challenges. SCs around the world are diverse in their characteristics and are facing unprecedented challenges. The pace of urbanization is rapidly increasing. Standards (such as established by ISO and IEEE) can play a crucial role in the development of SCs. We must ensure that the information is secure and the people are secure. Since networks are believed to be the least secure parts of the system, cities must ensure that the networks are safe before embarking on SC initiatives. Everyone is needed online and needs to be able to access services in order to realize the full benefits of IoT. Many US cities aim at becoming age-friendly because a society that is good for the elderly is good for everyone.

2.3 Smart Homes

Smart home is the main building block of SCs and intelligent communities. A home is a social space where family members interact. Modernizing existing homes by making them more efficient and "smart" offers enormous energy savings potential. The smart home is one of the emerging infrastructures. It is an application of ubiquitous computing that provides services to users in form of remote home control or home automation.

The term "smart home" refers to a residence that has all appliances (from cameras to coffee makers) that are capable of communicating with one another and can be controlled remotely. Smart homes are also called smart houses, intelligent homes, aware homes, or adaptive house. A smart home is usually a new building that is equipped with smart features and technology to enable occupants to remotely control devices in the home. A typical smart home is shown in Figure 2.3 [8].

There is a growing interest in the idea of smart homes. To be regarded smart, a home must include techniques of activity recognition and must be capable of providing a better quality of life for the inhabitants. It must be able to "think" for itself. Smart homes are supposed to enhance living experience and support independent living, especially for the elderly. Smart home technologies have been introduced all over the world.

The main objectives of a smart home are to improve the quality of life, increase automation, facilitate energy management, and minimize environmental emissions. Smart homes automate electrical devices and control the environment inside. A smart home allows the occupants to remotely arm

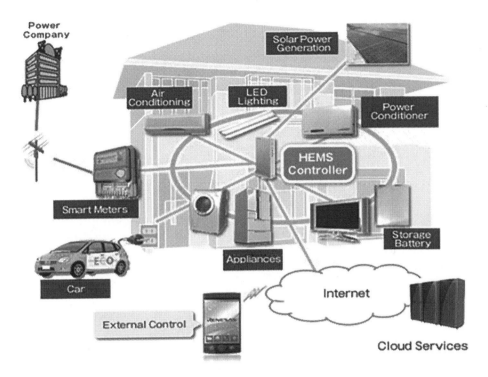

FIGURE 2.3
A typical smart home [8].

or disarm security, lock and unlock doors, control lights. They can also get alerts about events such as when children arrive home from school.

The concept has been extended to smart buildings such as smart offices, hospitals, and sport centers. The true benefits of the smart home will not be realized without improvement in underlying technologies.

2.3.1 Underlying Technologies

Technological innovations, from telecommunication to intelligent appliances, introduce new possibilities to the smart home. The underlying technologies for a typical smart home include the following [9]: (1) sensors, (2) computers, (3) software, (4) user input devices, (5) user output devices, (6) mechanical hardware, (7) wireless technology, (8) batteries and other power sources, (9) IoT, (10) smart appliances, (11) GPS, and (12) other related technologies. These can be grouped into three technologies: communication networks, sensor networks, and home automation technologies.

- *Communication networks.* Home devices are connected to form a home area network (HAN), where communications are enabled by different protocols. IoT will connect devices in the home to the

Internet. HAN, personal area network (PAN), neighborhood area network (NAN), and body area network (BAN) are communication infrastructures that are employed in domestic applications, allowing the user to be on the move [10]. The concept of smart homes promises that household devices communicate with their owners.

- *Sensor networks.* Since the smart home consists of a wealth of connected gadgets, it is equipped with different kinds of sensors: door sensors, motion sensors, temperature sensors, water flow sensors, accelerometer sensors, gas sensors, reed switch sensors, piezoelectric sensors, and light switch sensors [11]. These sensors, along with connected devices, can make home life more enjoyable, relaxing, fulfilling, and convenient. For example, sensors and monitors detect environmental conditions such as light, temperature, motion, and humidity. Wireless technology provides an invisible connection and communication among devices. The more bedroom gadgets one has, the better one might be able to sleep at night. With the gadgets one has in the kitchen, one feels inspired to cook and clean.

- *Home automation technologies.* A smart home has some measure of automation built into it. Smart home technology (also known as home automation) automatically monitors and controls water treatment, air conditioner, lighting, etc., and ensures that energy consumption is limited to what is required. It also provides homeowners security, convenience, and energy saving by allowing them to control smart appliances by networked devices, such as a smart phone. Home automation technologies include X10, ZigBee, Z-wave, Universal Powerline Bus (UPB), Smarthome, Arduino microcontroller, and Wi-Fi. These will let you control your home from anywhere you have access to the Internet.

2.3.2 Smart Home Devices

All of the electrical devices in a typical home can be made smart. Not just computers and smartphones, but everything: clocks, speakers, lights, doorbells, cameras, television, security systems, thermostats, sprinklers, intercoms, bulbs, thermostats, heaters, air conditioners, washing machines, windows, window blinds, stove, microwave, dishwasher, clothes washer, clothes dryer, hot water heaters, robot lawn mowers, appliances, cooking utensils, fridge, furnace, etc. This will lead to smart TV, smart meters, smart refrigerator, smart lightbulbs, smart thermostats, smart security, smart cameras, smart appliances, smart locks, and garage-door openers, etc. These smart devices are capable of communicating with each other and can be controlled remotely from any room in the home or from any location in the world.

2.3.3 Home Energy Management

Users are usually motivated by the cost-saving potential of smart homes. There are different kinds of users, such as home or apartment occupants, building management authorities, and office management authorities. Instead of depending on the power company, a smart home can generate its own energy using solar panels, wind turbines, geothermal plants, and other energy sources. Smart homes within the same neighborhood can form a microgrid and trade energy among themselves [12]. The energy price is dictated by the local microgrid and the utility provider which uses the Internet to receive energy consumption information.

Smart home energy can be managed by proper selection of efficient appliances. Smart meters can be employed for measuring energy consumption. A smart oven cooks faster and healthier based on selected recipe. It can communicate with other devices such as cell phone, smart fridge, and fire alarm. Managing power consumption with home automation helps in effective use of limited resources such as water, gas, and electricity.

2.3.4 Benefits and Challenges

One of the greatest motivations for having smart homes is to assist elderly and disabled people and enhance their well-being and independent living. Studies show that old people tend to isolate themselves from others, which may have a negative impact on their health, and that they prefer staying in their own homes, a phenomenon known as age-in-place.

Smart homes provide 24/7 monitoring that can help seniors to remain at home comfortably and safely [13]. With the worldwide development of aging societies, there is a calling for smart technologies providing independence, productivity, and quality of life among elderly people [14]. Senior citizens living in smart homes tend to live longer in their homes by reducing caregiver burden. Some senior citizens are not computer-literate. Some sensors are not appropriate for elderly people, who may not be willing to wear them. While people want smart domestic appliances to help in their menial tasks, they want to remain in control.

Smart homes can make life easier, more comfortable, and more convenient. They also provide some energy efficiency savings and allow being green. For example, a homeowner on vacation can use a smart phone, tablet, laptop, or Apple watch to interact and control a home security system, control temperature gauges, and switch appliances on or off. Not only will we be able to control our homes remotely, our home will be able to send us a message when they detect someone ringing the doorbell or be notified when family members return home. Other benefits include upgraded home appliances, safety mechanisms, centralized control, and remote access.

However, smart homes face some challenges and have not generated a breakthrough. Cost and work required for installation present major challenges. Many smart home opponents worry about security and data privacy risks with Internet-connected devices in homes. Security of a home is of paramount importance. If hackers succeed infiltrating a smart device, they could potentially turn off the lights and alarms and break into the home. Hacking is the most serious barrier to adoption of smart home technology. No one would like to compromise privacy. For privacy reason, one may decide to keep tech out of certain rooms, such as the bathroom or bedroom. Other challenges include consumers' ignorance of the benefits of smart homes, the hassle and cost of installing the smart devices, and complexity of the technology to most homeowners.

Smart homes are homes of the future. As prices for equipment fall and consumer demand rises, smart home technologies will make their way into more new homes. The benefits of a smart home (independence, safety, mobility, and energy cost savings) far outweigh the initial expense [15].

2.4 Smart Energy

Energy is a basic necessity in the modern world. The most common forms of energy are electricity, heating, cooling, solar, wind, and gas. Our discussion in this section will be mainly on smart energy to due electricity. To sustain our modern society, we need an uninterrupted supply of energy or electricity. Energy is the key to tackling the most important issues of today and tomorrow such as climate change, sustainable development, health and environment, global energy and food security, and environmental protection [16]. The world's annual electricity generation was 20,250 TWh in the year 2012 and is expected to be 25,500 TWh in the year 2020 [17]. Buildings consume a lot of energy and are responsible for the largest carbon dioxide (CO_2) producers. Therefore, making efficient use of energy at smart homes, SCs, and buildings is crucial for conservation and reduction in greenhouse effects.

The existing power grid provides one-way distribution of electricity from the power generator to consumers. The basic structure has not changed for about 100 years. It is known to be inefficient and unreliable. As a result of its low efficiency, the power industry is faced with unprecedented challenges and opportunities. Experiences have shown that it is not suitable for 21st century [18].

In recent years, the terms "smart energy" and "smart energy systems" have been used to express a broader approach than the term "smart grid." Where smart grids focus primarily on the electricity sector, smart energy includes of more sectors (electricity, heating, cooling, industry, and buildings) [19]. To be considered as smart, energy systems should use technologies and resources that are adequate, affordable, clean, and reliable. Smart energy systems are needed for SCs, smart homes, electric vehicles, smart meters, smart irrigation, etc. They are important component of IoT. Buildings account for more than 40% of energy consumption worldwide. Smart energy management has two major aspects, the smart grid (transmission and distribution system) on one hand and programmable smart appliances on the other.

The word "smart" in smart grid refers to the notion of a power grid with intelligence. A smart grid is commonly regarded as a digital upgrade of the existing power system.

The main objective of the smart grid is to bring reliability, flexibility, efficiency, and robustness to the power system. Smart grid does this by introducing two-way data communications into the power grid. It delivers electricity in a controlled, smart manner, from points of generation to active consumers. Thus, the smart grid consists of the power infrastructure and communication infrastructure, which correspond to the flow of power and information, respectively. This enables intelligent operation of the smart grid. But this introduces security-related challenges.

2.4.1 Components of a Smart Grid

The term "grid" is traditionally used for electricity generation, electricity transmission, electricity distribution, and electricity control. A "smart grid" is an enhancement of the traditional electric power grid. It is the modernization of the power delivery system. It is a transformation of the legacy unidirectional electric grid into automatic intelligent system of bidirectional exchange of electric power and information. A smart grid may be defined as any combination of enabling technologies, hardware, software, or practices that collectively make the delivery infrastructure (or the grid) more reliable, more versatile, more secure, more accommodating, more resilient, and ultimately more useful to consumers [20].

The consumers constitute an integral part of the smart grid because they are aware of energy use, can modify the purchasing patterns, and supply energy back into the grid. Power production technologies, such as photovoltaic cells and windmill plants, have given rise to "energy prosumers," users who produce and consume electrical energy at the same time.

A smart grid basically consists of overlaying the physical power system with the information system. A smart grid architecture is shown in Figure 2.4 [21]. From the technical point of view, the smart grid can be divided into three major systems [22]:

- *Smart infrastructure system.* This is the energy, information, and communication infrastructure underlying the smart grid. This allows two-way flow of electricity and information. This implies that the users may put back electricity into the grid. The system enables multiple entities (such as intelligent devices, dedicated software, control center, etc.) to interact.

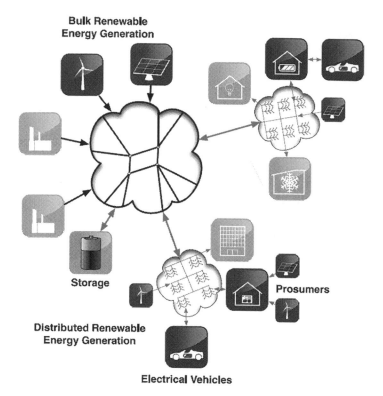

FIGURE 2.4
Smart grid architecture [21].

- *Smart management system.* This provides advanced management and control services. Efficient management is fundamental for efficient operation of smart grids. Management of smart grid includes the development and implementation of smart metering, real-time pricing, efficient management of renewable energy sources, and management of transmission and distribution networks.

- *Smart protection system.* This provides advanced reliability analysis, fault protection, and security services. The existing infrastructure has become vulnerable to several security threats.

- *SCADA.* Supervisory control and data acquisition (SCADA) systems are basically special computer-based networks and devices which are used in controlling and monitoring our critical infrastructures. There are 16 critical infrastructures in US. These include [23]: energy, waste and water systems, telecommunications, transportation, chemical, dams, emergency services, financial services, commercial facilities, government facilities, critical manufacturing, defense, food and agriculture, healthcare and public health, and nuclear reactor. These critical infrastructures are interdependent.

SCADA is a control system for smooth managing large-scale, automated industrial operations. One of the key capabilities of the smart grid results from the integration of SCADA. When applied to electric power industry, SCADA can help the industry to save time and money, reduce operational costs, and improve efficiency. It provides real-time monitoring and automation for smart power grid, a promising power delivery system of the future. It will continue to enable power companies to conduct their business in remote and hostile environments.

2.4.2 Smart Energy Management

Smart energy management has two major components: the smart energy system and smart meters.

1. *Smart energy system*. A smart energy system is affordable. It is based on three grid infrastructures [24]:
   - Smart electricity grids to connect flexible electricity demands such as heat pumps and electric vehicles to the intermittent renewable resources such as wind and solar power.
   - Smart thermal grids to connect the electricity and heating sectors.
   - Smart gas grids to connect the electricity, heating, and transport sectors.

   Smart energy grids can provide efficient bidirectional energy supply and enhance the operational efficiency of energy supply with reduced greenhouse gas emissions. They allow for intelligent monitoring and distributed energy generation capabilities within the multienergy systems (thermal, electricity, and gas, water). They facilitate the integration of diverse technologies such as renewable energy, electrical vehicles, and smart homes.

   The major expectations from smart energy systems are presented in Figure 2.5 [25]. An energy system is considered smart if it uses technologies and resources that are adequate, affordable, clean, and reliable. Therefore, smart energy systems are evaluated based on their efficiencies and environmental performance. The successful functioning of smart energy systems necessitates strong minded and direct action.

2. *Smart meters*. These are electronic devices that are used in a home or business to measure how much energy is consumed. They are the most basic components in the intelligent energy networks. Although smart metering technologies have so far been mainly used electric smart grids, recent development has enabled auto reading and two-way communications of heat and gas meters [26]. Smart meters bring an end to home visits from meter readers because they can remotely

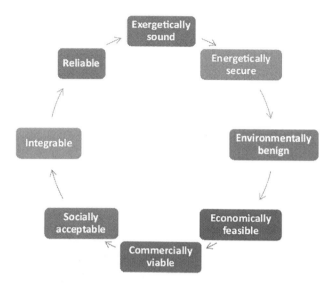

FIGURE 2.5
Major expectations from smart energy systems [25].

record and report utility consumption (energy, gas, or water). They provide real-time information necessary to prevent malfunctions and damages to utilities.

Other technologies involved in smart energy management include IoT and RFID.

2.4.3 Benefits and Challenges

Smart energy has the potential to reduce energy bills of households and businesses. It also reduces the demand for oil and gas and creates new green jobs. While smart energy technology has positive impacts on global warming, health, and cost, it has negative effects on security and privacy [27].

Data security and privacy become important when IoT technologies are employed for smart energy applications. Smart meter invites intended or unintended privacy breaching activities such as in-house activity detection. Careful planning is required to ensure that all possible measures are taken to prevent compromise. Energy security is also important for the advancement and improvement of all societies [26].

Smart energy grids present huge technical challenges in their design, operation, and maintenance. They also present challenges for communication networks and information technologies. Other key benefits of smart grid include uninterrupted supply of power, reduced transmission and distribution loss, secure grid, and market based electricity pricing [27,28].

2.4.4 Security

As we move from legacy power systems to more modern smart grid systems, security will be a big issue. For any new system, security has to be part of system design.

Security is the key factor in system design at each level of smart grid from metering to remote sensing and control networks. The security objectives have been availability, authentication, confidentiality, nonrepudiation, and integrity. Security threats within the smart grid usually attempt to compromise one or more security objectives. As the smart grid depends on computer networks, it is vulnerable to various security and privacy issues. A cyber attacker can penetrate a smart grid using variety of attacks. The physical layer security for smart grid deals with unauthorized access, malicious attacks, privacy issues, and voltage regulation. To secure a smart grid, it is important to have several mechanisms in place. These include authentication protocols, cryptographic algorithms, and firewalls [29].

The implementation of smart energy systems will be a long continuous process because it involves technological and financial investment. It also involves international effort. The government of each nation will need to develop a policy for implementing smart grid. As the smart grid moves in people's living room, the focus will significantly change to marketing to consumers. A good source of information about smart grid is the *IEEE Transactions on Smart Grid* (2010–present).

2.5 Smart Transportation

Transportation is an important factor that affects the quality of life. Its development is regarded as the most significant driver for social progress. Transport network connects cities, nations, manufacturers, and retailers. Transportation enables mobility, allows people to interact, and facilitates the growth of economy since it enables delivery of goods and services around the world. Modes of transportation include air, sea, and land.

Although transportation has improved our lives, a number of costly problems remain unsolved, including traffic accident, congestion, and vehicle emission.

The department of transportation will never have enough financial resources to supply the endlessly increasing transportation demand by building more, wider, and faster state roadways. "To accommodate increasing transportation demands and provide safe and efficient travel in U.S. communities, it is estimated the nation's highway and bridges will need $290 billion in investments, transit systems $18 billion and airports $50 billion over the next 5 years" [30]. A number of cities have started using smart transportation as a solution to transportation-related problems.

Smart transportation is an important component of the SC. It refers to the application of ICT to road transport, infrastructure, vehicles, and users. It improves safety, increases productivity, and ensures greener environment. Smart transportation (and its cousin, intelligent transportation system) offers a means of providing innovative services on different modes of transportation and traffic management. It is an important area in the smart grid, SCs, and smart vehicles. Its components include infrastructure, vehicles, and users [31]. By nature, smart transportation applications are examples of vehicular ad hoc network (VANET) applications.

2.5.1 Enabling Technologies

In general, the transportation system consists of physical and organizational objects (such as ICT, transportation infrastructure, vehicles) interacting with each other to enable smart transportation and logistics activities. Smart transportation is enabled by a number of technologies:

- *IoT.* Smart transportation is being influenced by communications networks and the IoT. IoT connects billions of smart devices. Connected smart cars, buses, trains, and air planes will allow people to always stay connected to the Internet. However, IoT privacy and security concerns are serious.

- *Wireless technologies.* Several wireless technologies have been proposed to smart transportation, but there is lack of consensus as to which technology is the best [32]. Smart mobile devices are used in route planning, navigation or road guidance (GPS), carpooling, and parking information. Cellular phone systems, such as Wi-Fi, and Bluetooth, create a field of data connectivity.

- *Sensing technologies.* With sensors, RFID, and other connected technologies, it is feasible to connect everything (traffic lights, road signs, etc.). These sensors are used in sensor-enabled consumer devices.

- *GPS.* A navigation system such as global positioning system (GPS) allows the user to find the best route based on real-time conditions. More and more vehicles have in-built GPS (satellite navigation) systems. But Google maps and similar navigation systems are replacing built-in GPS systems. Two smart cars will override the drivers and avoid collision if they know where they are. 3D camera-enabled dash board navigation device can be used to enhance accident prevention.

As shown in Figure 2.6, smart transportation can be ushered in through improvements in four major areas: smart automobiles, smart infrastructure,

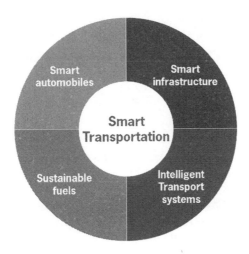

FIGURE 2.6
Smart transportation [33].

intelligent transport system (ITS), and sustainable fuel [33]. ITS refers to the integrated application of modern technologies and management strategies in our surface transportation systems. It uses existing infrastructure more efficiently and avoids the need to build more roads. Although ITS will not solve all transportation problems, by combining technological advances with the existing transportation system, we can build smarter highways, smarter traffic lights, smarter parking, and smarter vehicles.

2.5.2 Applications

Smart transportation applications are often large and complex since they need to handle heterogeneous, dynamic devices such as mobile vehicles and roadside units. The following are common application areas for smart, intelligent transportation.

- *SCs.* Applications of smart transportation services have led to the rapid growth in SCs all over the world. The goal of SCs is to achieve high quality of life through the use of technology and environmentally friendly practices. The users are empowered to decide where and when they want to travel.

- *Smart vehicles.* Smart vehicles serve as nodes in the IoT as part of smart transportation. Transportation infrastructure (consisting of smart traffic lights and camera network) can help autonomous vehicles find their way. Semi-autonomous features like lane departure warnings already available in cars can reduce traffic accidents, while fully autonomous vehicles may reduce them even further.

Vehicle-to-vehicle communications will enable autonomous vehicles to communicate in real time.

- *Electric vehicles.* Electric vehicles need power charging infrastructure which will compose of charging stations and diverse vehicles. Battery-charged power sources may serve as back-up power source. Drivers can charge their vehicles at homes, offices, commercial charging stations, universities, shopping malls, airport parking lots, and the like [34].

- *Smart parking.* Parking is often difficult in a large city. Smart parking is usually regarded as a part of SC. Searching for parking spot can waste time and gas and might also cause traffic congestion and emissions. Smart solutions optimize the use of parking lots by equipping each parking space with a sensor which detects whether a car is parked there or not. Smart parking uses the technology of IoT and mobile payment to optimize the parking experience.

- *Emergency service network.* Globally, governments have realized the need of efficient Emergency Service Network (ESN) in case of emergency. The police, fire-fighting, and ambulance, must always have the highest priority for accessing reliable network services to save lives and avoid more loss. The main advantage of using such a dedicated network is that it enables the absolute usage of network for the ESN responsive teams [35].

2.5.3 Benefits and Challenges

Smart transportation has the potential to increase sustainability of moving people and goods from one location to another. It has a major impact on the citizen's life and the economy of the city. It will transform and automate our roadways, railways, and airways.

Implementing smart transportation will save money, increase safety, reduce gas emissions, increase efficiency, and provide a better quality of life. It will also increase the sustainability of how goods and people migrate from one place to another [36]. It will have an impact on our social and economic life. It will address all aspects of transportation including smart driving, smart parking, smart rapid transit, pollution, and security system. Smart transportation will provide higher productivity and efficiency of whole transport system, fewer traffic accidents, less commute time, more environmental friendliness, better passenger experience, and better life quality [37].

Yet the technology is not without its challenges. Critics against implementing smart transportation highlight some challenges: privacy protection, security, regulation, and sustainability. Due to security, many users/drivers are hesitant to provide their credit card information to an app-based transportation service provider. Security experts claim that the system is vulnerable to attacks and hacks. This may endanger public safety [38].

Since resources are limited, transportation problems will always outweigh available resources. Awareness about smart transportation is still low in developing countries, where the infrastructure is the main obstacle for implementing smart transportation [39].

In spite of these challenges, several cities have started implementing smart transportation. Smart transportation must accommodate existing transportation infrastructure and facilitate transition from the existing to future structure. The global smart transportation market will keep growing.

The Smart Transportation Alliance (STA) is a not-for-profit global organization that was founded in 2014 with headquarters in London, UK. Its mission is to lead and support activities improving the methods, technologies, and standards associated with smart transportation.

2.6 Smart Agriculture

Agriculture is the oldest sector of the US economy. For most people living in the developing countries such as in Africa and South Asia, agriculture is the major employment source. Today, agriculture industry faces a number of great challenges in an attempt to feed the 9.6 billion people that the UN Food and Agriculture Organization (FAO) predicts are going to be living on earth by the year 2050. These challenges include [40]: (1) increasing cost of farming; (2) farmlands have been replaced by factories or houses due to urbanization and population explosion; (3) stress on water resources due to chemical pollution; (4) food wastage due to consumers wasting food; (5) complying to government regulations adds to unreasonable costs; (6) farms are becoming more like factories: tightly controlled operations for turning out reliable products; and (7) agriculture causes more than one fifth of the global emission of greenhouse gases.

The impact of climate change occurs at the regional, national, and global scales. Climate change affects crop product and hinders agricultural growth in several parts of the world. It affects rainfall. It also poses a threat to livestock production. It calls for collective action. National governments and corporate sectors can provide coordinated approaches to climate change, integrated risk management, agricultural, and food security policies.

Smart agriculture can be described as the use of different advanced technologies toward in the agriculture domain. It is variably regarded as smart farming, high-tech farming, precision agriculture, or smart-climate agriculture (SCA). It represents the combined application of ICT solutions such as precision equipment, and the IoT. IoT and smart devices are helping farmers to remotely monitor equipment, crops, and livestock. This is leading to what is called a Third Green Revolution. "The first American agricultural revolution in the 1860s was characterized by the change from hand power to horses and the second American agricultural revolution in the 1940s was characterized by the change from horses to tractors, we are definitely entering a Third American Agricultural Revolution with Smart Farming or Precision Agriculture powered by Technology" [40]. The US currently leads the world in the application of IoT in smart agriculture.

2.6.1 Smart-Climate Agriculture

SCA avoids the lose–lose situation by integrating climate change in agriculture strategies. The term was coined by the UN FAO in 2010. SCA promises to transform agricultural systems that will decrease global food insecurity and reduce poverty. SCA practices can raise farm productivity while mitigating climate change. By promoting new methods and technologies, SCA helps farmers to manage their resources, boost their profits, and reduce agriculture's contribution to climate change. Even small-scale farmers in developing nations can achieve success and increase farm production by adopting SCA technologies.

Productivity (or food security), adaptation, and mitigation are the three interrelated pillars for achieving SCA. SCA plans to increase agricultural productivity without making a negative impact on the environment. It aims to build farmers' capacity to adapt and prospect in the face of odds. It helps to reduce greenhouse emissions. It facilitates climate-change adaptation for farmers.

In contrast to traditional agriculture, SCA integrates climate change and agricultural development. SCA may involve a wide range of technological innovations, water management, and agro-forestry. Adopting it at farm scale may be influenced by institutional mechanisms, landscape governance, socioeconomic factors, and climate conditions. To achieve climate-change objectives, agricultural systems must become climate-smart landscapes. This involves integrating agricultural landscape management with adaptation and mitigation [41].

2.6.2 Smart Farming

Smart farming refers to the adoption of ICT to enhance and automate agricultural processes and operations. It is a farming management concept that involves using modern technology to increase the quantity and quality of agricultural products. Smart farming technologies cover all these aspects of precision agriculture. It applies to small family farming and complex family faming as well as organic farming.

According to the UN FAO, the world will need to produce 70% more food in 2050 than it did in 2006. Smart farming addresses this challenge of feeding the ever-increasing world population. It is a high-tech, capital-intensive means of sustainably growing food for the masses. Smart farming is the adoption of ICT in order to enhance, monitor, and automate agricultural processes.

2.6.3 Enabling Technologies

Combining emerging technologies, such as ubiquitous computing, context-aware computing, and cloud computing, with wireless sensor networks can be applied to make the agriculture smarter. Farmers use technologies such as mobile phones, robots, and computers along with tractors and hoes to produce more with less. They produce magical results that can transform the entire agriculture industry. The basic technologies for smart farming include precision equipment, the IoT, sensors and actuators, GPS, BD, smart phones, unmanned aerial vehicles (UAVs, drones), and robotics (agrobs). Some of these technologies are illustrated in Figure 2.7 [42] and explained below.

- *The IoT.* The application of IoT in agriculture can have a great impact. IoT can address many challenges and increase the quality, quantity, and cost-effectiveness of agricultural production. The IoT is posed to push the future of farming to the next level. It will assist farmers in meeting the world's food demands in the coming years.
- *Sensor network.* Although this is part of IoT technologies, it needs to be treated on its own. With the ICT, it is possible to create a sensor network that can be used for continuous monitoring of the farm [43]. Sensors watch everything—temperature, humidity, illumination. Cattle, for example, are getting their own private sensors. Sensors placed in the farm are used in collecting information (such as soil

FIGURE 2.7
Technologies involved in smart farming [42].

moisture, fertilization, temperature, humidity, acidity, illumination) and providing farmers with real-time data on their land, crop, livestock, and equipment. Farmers can use data collected from remote sensors to decide where their water resources should be channeled. In 2001, John Deere, the world's largest agricultural equipment manufacturer, decided to fit their tractors and other machines with GPS sensors. Sensors attached to moving machinery can be used to take measurements on the run. Sensors can be placed inside cattle to measure stomach acidity and examine digestive problems.

- *Agricultural automation.* Autonomous, robotic vehicles, and artificial intelligence techniques have been developed at all levels of agricultural production such as mechanical weeding, applying fertilizer, or harvesting of crops. Drones and drone-mounted cameras are used by farmers to survey their lands. GPS-guided tractors are becoming the norm. To get a synoptic view of their farm, some farmers employ satellites. These technical improvements constitute a revolution that will cause disruptive changes in agricultural practices.

2.6.4 Benefits and Challenges

One of the most important advantages of smart agriculture is the ability to grow more food with fewer resources. Smart agriculture has a real potential to deliver a more productive and sustainable agricultural production, based on a more precise and resource-efficient approach. It reduces human errors and improves yields. It enables intelligent, automated irrigation system which uniformly provides water to the whole field.

However, adoption of CSA or smart agriculture faces a number of challenges. Choosing the right type of IoT technologies to use can be intimidating to farmers since there are hundreds of them. The massive energy consumption of IoT devices can create an impact to the environment. Smart agriculture is relatively new and lacks empirical evidence.

There is a growing divide between how developed and developing countries embrace smart agriculture practices and technologies. This divide has limited the potential benefit of smart agriculture to integrate adaptation, mitigation, and food security outcomes [44]. There is lack of knowledge or information to the farmers.

Socioeconomic factors such as culture, education, poverty, and investment costs may limit the widespread adoption of smart agriculture. Although various tools and equipment with the latest technology can be used in the farm, most farmers cannot afford them. The cost of many labor-saving technologies is prohibitive.

Some farmers have poor training in technology and are often skeptical of change since it can be painful and costly to get things wrong. Limited knowledge and technical skills of individual farmers can be major hurdles particularly in developing nations. There is currently no single policy approach for achieving smart farming. A proactive development of policies for smart farming supporting the necessary legal and market architecture is necessary. For a smart farm to be fully automatic and factory-like, people would have to be cut out of the loop altogether. This requires introducing robots on the ground as well as in the air. Constant climate change, soil degradation, and water shortages are also hurdles that continue to increase in size with time. Small changes in these factors can have a major impact on profitability.

Adaptation in agriculture sector is necessary for eradicating poverty and hunger. Adoption of smart agriculture practices by farmers has been low globally despite its benefits. Shareholders must decide the appropriate policies and practices toward a viable agricultural production system. Governments must ensure smart agriculture practices and technologies are integrated in their economic development strategies. Smart farming is the future of agriculture [45].

2.7 Smart University

Today's universities and university professors widely adopt digital technologies and can be regarded as already having the desired smartness characteristics. However, they can become "smarter" to improve their effectiveness, enhance their performances, and be more flexible to meet the requirements of the labor market and smart society. All universities should make efforts to become smarter in order to optimize the teaching–learning process.

The use of digital resources has drastically changed the way individuals interact with their university environment. Information and communication devices are increasingly employed in university teaching–learning environment to improve the quality of education and individuals' cultural life. Today's universities are widely adopting digital technologies and can be regarded as already having the desired smartness characteristics.

Smart university is an emerging concept with a new ubiquitous computing and communication field. It will radically change the education system by providing devices supported with smart technologies. It will enhance communication among students, professors, and administrators. The smart university model is being developed to promote modernization of universities all over the world.

The main objective of the smart university is to enhance the quality of educational, research, commercial, social and other activities of the university. This will help meet the requirements of the labor market and smart society where the use of smart technologies, cloud computing, and open innovations is common. Smart university may be regarded as a component of smart and sustainable city development.

2.7.1 Features of Smart University

The main features of the conventional university include infrastructure (buildings, classrooms, parking spaces, and laboratories), education, human resources, and management. All these need digitization to make the university smart. The smart university is a system based on a smart campus (or smart learning environments), smart classrooms, smart faculty, smart students (or creative students), smart pedagogy, smart curriculum, smart technologies and services, and other smart distinctive features. Essentially, a smart university consists of five entities: smart people, smart building, smart environment, smart governance, and knowledge grid, as illustrated in Figure 2.8 [46].

- *Smart education.* Education is considered smart when it is provided in a smart environment supported by smart technologies, making use of smart tools and smart devices. Smart education is transforming conventional way of pedagogy to contemporary techniques using ICT. This harnesses potentials of technologies, pedagogies, and digital textbooks to change education systems. It creates unique opportunities for academic institutions and organizations.

- *Smart classroom.* Smart classroom uses collaborative teaching and the latest technological teaching tools to create a modern and effective education environment. It creates a best practice standard for implementing modern, academic curricula. This makes learning interesting through the use of digital devices such as VCD or DVD player, laptops, computers, and projector. It will also involve technologies

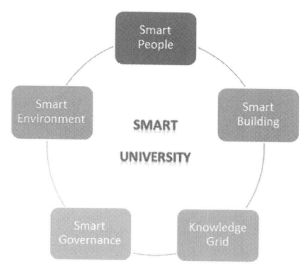

FIGURE 2.8
Components of a smart university [46].

such as smart boards, smart screens, wireless keyboard, wireless mouse, and wireless Internet access, which may be provided by Wi-Fi and ZigBee. Such a smart learning environment enables learners to access digital resources and interact with learning systems in any place and at any time.

- *Smart campus.* This is like a small world where sensor-enabled devices are interconnected to make living more comfortable to its inhabitants. It is a mini SC. As a learning environment supported by smart technologies, a smart campus (or intelligent campus) will offer instant and adaptive support to learners. This may involve ICT sustainability with intelligent sensor management systems and smart building management with automated security control and surveillance.

- *Smart governance.* The smart university can adapt and accommodate various types of students such as regular students and life-long learners, in-classroom or distance/online students, students with special needs such as students with disabilities. A university becomes smart when [47]: (1) it contains less bureaucracy, (2) it facilitates the creation of smart students, and (3) it supports innovative investments fostering synergy between teaching and research. This will also involve smart enrollment, smart staff and resource management, and course management.

- *Smart people.* Students these days are more technological than ever and are demanding innovative ways to learn. A smart university employs smart teachers and gives them smart tools while assessing their pedagogical effectiveness using smart evaluation process. Smart education means flexible learning in an interactive learning environment and making it available to everybody, everywhere, anytime.

2.7.2 Enabling Technologies

Smart universities will use social networking services (such as Facebook) to facilitate online communication and collaboration among students. In addition, digital technologies enabling smart universities include the following:

- *ICT.* Smart university may be regarded as an ICT-driven university. ICTs enable interconnectedness, which is needed for the delivery of smart education [48]. IoT and wireless sensor network (WSN) can be used for building a smart university.

- *RFID.* The radio-frequency identification (RFID) system consists of tags (transponders) and tag reader. It is used in taking care of maintaining attendance record, switching control of electrical items, and security locks of rooms. It can contribute to improving security, power conservation, person tracking, etc. For the purpose of smart identification, each student, employee, office, classroom, and lab should have a unique ID. RFID should benefit teachers and students with identification, tracking, smart classroom, smart lab, smart attendance recording, etc.

- *NFC.* The near-field communication (NFC) is an emerging technology which is suitable for the implementation of the smart university. NFC is based on RFID and it combines wireless proximity communication with mobile devices. The smart university model is a ubiquitous computing platform based on NFC devices which interact with each other to create an intelligent environment. The technology allows the interaction in the environment by bringing the mobile device into contact with smart objects by just "touching" them with their NFC-enabled devices such as mobile phones.

2.7.3 Benefits and Challenges

Smart university is an integral part of the innovation toward SC. It is the wave of the future in a highly technological society. Some of the advantages of a smart university include: reduction in electricity consumption, creating a conductive environment for socialization, easy to report all irregularities, easy location of people on campus, avoiding stealing of equipment, and ease maintenance of attendance record. Students can create their own online learning activities, while parents can observe and support their child's learning from home.

Although technology can make learning interesting and entertaining, the use of electronic gadgets can impair the problem solving capacity of learners. In most cases, students' parents are a generation behind and may not be able to support and assist them as they would like to.

2.8 Conclusion

The large deployment of the IoT is enabling smart everything.

Building IoT framework in cities and homes creates intelligent infrastructures and devices, which are crucial for the future smart world. IoT is also deployed in creating smart energy or smart power grid, in making existing transportation infrastructure smart, and in helping smart agriculture. When smart infrastructures fail, it can lead to serious disruption of society and economy. The challenge is to make smart solutions resilient.

References

1. T. K. L. Hui, R. S. Sherratt, and D. D. Sanchez, "Major requirements for building smart homes in smart cities based on Internet of Things technologies," *Future Generation Computer Systems*, vol. 76, 2017, pp. 358–369.
2. A. Zanella et al., "Internet of Things for smart cities." *IEEE Internet of Things Journal*, vol. 1, no. 1, February 2014, pp. 22–32.
3. T. H. Kim, C. Ramos, and S. Mohammed, "Smart city and IoT," *Future Generation Computer Systems*, vol. 76, 2017, pp. 159–162.
4. M. N. O. Sadiku, A. E. Shadare, E. Dada, and S. M. Musa, "Smart cities," *International Journal of Scientific Engineering and Applied Science*, vol. 2, no. 10, October 2016, pp. 41–44.
5. R. Khatoun and S. Zeadally, "Smart cities: concepts, architectures, research opportunities," *Communications of the ACM*, vol. 59, no. 8, August 2016, pp. 46–57.
6. S. B. Letaifa, "How to strategize smart cities: revealing the SMART model," *Journal of Business Research*, vol. 68, 2015, pp. 1414–1419.
7. M. N. O. Sadiku, S. M. Musa, and O. D. Momoh, "Open government," *International Journal of Engineering Research and Advanced Technology*, vol. 2, no.12, December 2016, pp. 17–20.
8. "Smart home," http://internetofthingsagenda.techtarget.com/definition/smart-home-or-building.
9. W. C. Mann and B. R. Milton, "Home automation and smart homes to support independence," in W. C. Mann (ed.), *Smart Technology for Aging, Disability, and Independence*. Hoboken, NJ: John Wiley & Sons, 2005, pp. 33–66.
10. T. D. P. Mendes et al., "Smart home communication technologies and applications: wireless protocol assessment for home area network resources," *Energies*, vol. 8, 2015, pp. 7279–7311.
11. A. Benmansour, A. Bouchachia, and M. Feham, "Multioccupant activity recognition in pervasive smart home environments," *ACM Computing Surveys*, vol. 48, no.3, December 2015, pp. 34.
12. B. Zhang, P. P. Rau, and G. Salvendy, "Design and evaluation of smart home user interface: effects of age, tasks and intelligence level," *Behavior & Information Technology*, vol. 28, no. 3, May–June 2009, pp. 239–249.
13. G. Demiris and B. Hensel, " 'Smart homes' for patients at the end of life," *Journal of Housing for the Elderly*, vol. 23, no. 1–2, pp. 106–115.
14. M. R. Alam, M. St-Hilaire, and T. Kunz, "Computational methods for residential energy cost optimization in smart grids: a survey," *ACM Computing Surveys*, vol. 49, no. 1, April 2016, pp. 2.1–2.34.
15. M. N. O. Sadiku, S. M. Musa, and R. Nelatury, "Smart homes," *Journal of Scientific and Engineering Research*, vol. 3, no. 6, 2016, pp. 465–467.
16. I. Dincer and C. Acar, "Smart energy systems for a sustainable future," *Applied Energy*, vol. 194, 2017, pp. 225–235.
17. B. M. Buchholz and Z. Stycznski, *Smart Grids – Fundamentals and Technologies in Electricity Networks*. Heidelberg: Springer-Verlag, 2014, p. 19.
18. M. N. O. Sadiku, S. M. Musa, and S. R. Nelatury, "Smart grid – An introduction," *International Journal of Electrical Engineering & Technology*, vol. 7, no.1, January–February 2016, pp. 45–49.

19. H. Lund et al., "Smart energy and smart energy systems," *Energy*, vol. 137, 2017, pp. 556–565.
20. F. P. Sioshansi (ed.), *Smart Grid: Integrating Renewable, Distributed, and Efficient Energy*. Oxford, UK: Academic Press, 2012, pp. xxix, xxx, 393.
21. H. A. H. Hassan, A. Pelov, and L. Nuaymi, "Integrating cellular networks, smart grid, and renewable energy: analysis, architecture, and challenges," *IEEE Access*, vol. 3, 2015, pp. 2755–2770.
22. X. Fang et al., "Smart grid – the new and improved power grid: a survey," *IEEE Communications Survey and Tutorials*, vol. 14, no. 4, Fourth Quarter, 2012, pp. 944–980.
23. L. Goike, "Susceptibility of SCADA systems and the energy sector," *Masters Thesis*, Utica College, December 2015.
24. Q. Sun et al., "A comprehensive review of smart energy meters in intelligent energy networks," *IEEE Internet of Things Journal*, vol. 3, no. 4, August 2016, pp. 464–479.
25. I. Dincer and C. Acar, "Smart energy systems for a sustainable future," *Applied Energy*, vol. 194, 2017, pp. 225–235.
26. K. T. Raimi and A. R. Carrico, "Understanding and beliefs about smart energy technology," *Energy Research & Social Science*, vol. 12, 2016, pp. 68–74.
27. J. Momoh, *Smart Grid Fundamentals of Design and Analysis*. Hoboken, NJ: John Wiley & Sons, 2012, p. 1, 130.
28. J. Hastings, D. M. Laverty, and D. J. Morrow, "Securing the smart grid," *Proceedings of Power Engineering Conference (UPEC)*, 2014.
29. M. N. O. Sadiku, M. Tembely, and S. M. Musa, "Smart grid cybersecurity," *Journal of Multidisciplinary Engineering Science and Technology*, vol. 3, no. 9, September 2016, pp. 5574–5576.
30. D. Levy, "Getting smarter about transportation," *The American City & County*, vol. 112, no. 4, April 1997, pp. 28–30.
31. M. N. O. Sadiku, A. E. Shadare, and S. M. Musa, "Smart transportation: a primer," *International Journal of Advanced Research in Computer Science and Software Engineering*, vol. 7, no. 3, March 2017, pp. 6–7.
32. "Intelligent transportation system," *Wikipedia*, the free encyclopedia, https://en.wikipedia.org/wiki/Intelligent_transportation_system.
33. "Smart transportation – transforming Indian cities," www.grantthornton.in/globalassets/1.-member-firms/india/assets/pdfs/smart-transportation-report.pdf.
34. H. J. Kim et al., "An efficient scheduling scheme on charging stations for smart transportation," in T. Kim, A. Stoica, and R.-S. Chang (eds.), *Security-Enriched Urban Computing and Smart Grid*. Communications in Computer and Information Science, vol. 78, 2010, pp. 274–278.
35. "Smart transportation: maximize mobile network's value beyond connectivity," White paper, Feb. 2017, www.huawei.com/en/xlabs/insights-whitepapers/smart-transportation.
36. L. Schewel and D. M. Kammen, "Smart transportation: synergizing electrified vehicles and mobile information systems," *Environment: Science and Policy for Sustainable Development*, vol. 52, no. 5, September/October 2010, pp. 24–35.
37. "Smart transportation: maximize mobile network's value beyond connectivity," www.huawei.com/en/industry-insights/mbb-2020/trends-insights/smart_transportation.

38. S. Greengard, "Smart transportation network drive gains," *Communications of the ACM*, vol. 58, no. 1, January 2015, pp. 25–27.
39. A. Tedjasaputra and E. Sari, "Sharing economy is smart city transportation services," *Proceedings of the SEACHI 2016 on Smart Cities for Better Living with HCI and UX*, May 2016, pp. 32–35.
40. "Successful farming with technology—the new era!" https://medium.com/@Unfoldlabs/successful-farming-with-technology-the-new-era-e73cf4fc0dce.
41. M. N. O. Sadiku, M. Tembely, and S. M. Musa, "Climate-smart agriculture," *International Journal of Advanced Research in Computer Science and Software Engineering*, vol. 7, no. 2, February 2017, pp. 148–149.
42. "Toward smart farming: agriculture embracing IoT vision," www.google.com/search?q=smart+farming&tbm=isch&tbo=u&source=univ&sa=X&ved=0ahUKEwjyku7_2oPaAhUs0YMKHVXXCTgQsAQIYg&biw=2560&bih=1283#imgrc=1U__IX-hGPxMMM:&spf=1521851701360.
43. A. Walter et al., "Opinion: smart farming is key to developing sustainable agriculture," *Proceedings of the National Academy of Sciences of the USA*, vol. 114, no. 24, 2017, pp. 6148–6150.
44. A. Chandra et al., "Resolving the UNFCCC divide on climate-smart agriculture," *Carbon Management*, vol. 7, no. 5–6, 2016, pp. 295–299.
45. F. Tongke, "Smart agriculture based on cloud computing and IOT," *Journal of Convergence Information Technology*, vol. 8, no. 2, January 2013, (7 pages).
46. M. Owoc and K. Marciniak, "Knowledge management as foundation of smart university," *Proceedings of the Federated Conference on Computer Science and Information Systems*, 2013, pp. 1267–1272.
47. M. Coccoli et al., "Smarter universities: A vision for the fast changing digital era," *Journal of Visual Languages and Computing*, vol. 2, 2014, pp. 1003–1011.
48. S. I. Popoola et al., "Smart campus: data on energy consumption in an ICT-driven university," *Data in Brief*, vol. 16, 2018, pp. 780–793.

3

Big Data

Technology is dominated by two types of people: those who understand what they do not manage, and those who manage what they do not understand.

Putt's law

3.1 Introduction

We are in the information age and there are data everywhere. Each day, large amounts of data are being generated. Organizations and companies now have very large data sets stored in their files, databases, and data warehouses. They can access such data and extract new ideas about their products, customers, and markets. This increases innovation, value creation, and competitive advantage [1].

"Big Data" (BD) is an emerging term used with business, engineering, and other domains. The ability to collect and analyze huge amounts of data is a growing problem within the engineering community. BD is a high-volume, high-velocity, and high-variety information that requires special information processing tools [2]. It is also an area that requires multidisciplinary collaboration, as shown in Figure 3.1 [3]. It requires a new kind of professional: the "data scientist."

The birth of the concept of BD is usually associated with an META Group report by Doug Laney [4]. BD may be regarded as the result of an evolutionary process in IT that started in the 1960s. Two major factors are responsible for the start of the BD era. First, there was an increase in socialization with the advent of social media like Facebook, Twitter, LinkedIn, and Instagram. Second, cloud computing created the possibility of accessing and storing data over the Internet instead of on the hard drive of a personal computer [5].

This chapter presents the fundamental concepts of BD. It begins by describing what BD is all about and how it is generated. Then it presents wireless BD, big data analytics (BDA), and common tools for analyzing BD. The chapter further discusses typical applications of BD which include smart power grid, Internet of Things (IoT), business, industry, healthcare, nursing,

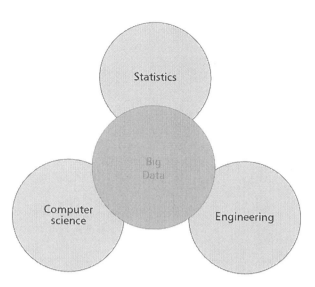

FIGURE 3.1
Big data requires multidisciplinary approach [3].

business, manufacturing, and public agencies. Then it addresses various benefits and challenges of BD.

3.2 Characteristics of BD

BD applies to data sets of extreme size (e.g., exabytes, zettabytes) which are beyond the capability of the commonly used software tools. It generally refers to the following types of data: (1) enterprise data, (2) machine-generated/sensor data, and (3) social data. It may require different strategies and tools for profiling, measurement, assessment, and processing. Typical sources of BD are shown in Figure 3.2 [6].

Originally, BD was defined to have only three basic properties: volume, velocity, and variety. This has been improved to include veracity and value. So we now characterize BD by the 5Vs: volume, velocity, variety, veracity, and value, as shown in Figure 3.3 [6a].

- *Volume.* This refers to the size of the data being generated both inside and outside organizations and is increasing annually. The sources may include text, audio, video, images, social networking, medical data, weather forecasting, etc. BD refers to data sets of extreme size (e.g., exabytes, zettabytes) which are beyond the capability of the commonly used software tools.

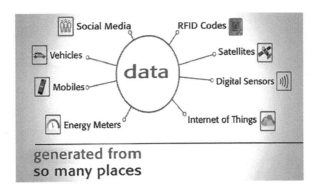

FIGURE 3.2
Sources of big data [6].

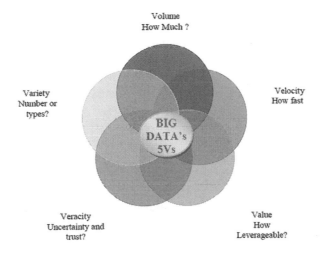

FIGURE 3.3
The 5V of big data [6a].

- *Velocity.* This refers to the speed at which data are generated by Internet users, mobile users, social media, etc. Data are generated and processed in a fast way to extract useful, relevant information. BD could be analyzed in real time, and it has movement and velocity.

- *Variety.* This refers to the fact that BD originates from heterogeneous sources and is in different formats (e.g., videos, text, logs). BD comprises of structured, semi-structured, or unstructured data. Structured data (such as bank transactions) are easy to analyze and validate using structured Queries. Unstructured data (such as email, text, images, etc., originating from social networking sites) are difficult to handle. Most data are unstructured. Keep in mind that what is structured or unstructured is not the formatting but the content of the document.

- *Veracity.* By this, we mean the truthfulness of data, i.e., whether the data come from a reputable, trustworthy, authentic, and accountable source. It suggests the inconsistency in the quality of different sources of BD. The data may not be 100% correct. In other words, how certain are we about the data? How good and clean are the data for performing analysis? Are the data from a reputable and trustworthy origin? The more data we accumulate, the harder these become to keep everything correct and consistent.

- *Value.* This is the most important aspect of the BD. It is the desired outcome of BD processing. It refers to the process of discovering hidden values from large datasets. It denotes the value derived from the analysis of the existing data. If one cannot extract some business value from the data, there is no use managing and storing it.

On this basis, small data can be regarded as having low volume, low velocity, low variety, low veracity, and low value. BD involves high volume, high velocity, high variety, high veracity, and high value, creating a situation where the data are too big, too fast, or do not fit the traditional database architecture. BD processing includes data collection, preprocessing, storage and management, analysis and mining, visualization, and application.

3.3 Sources of BD

BD refers to massive amount of data that are so large that traditional processing tools cannot cope. BD sources can be divided into three categories: (1) from human activities, (2) from the physical world, and (3) from computers. It is generated by infrastructures designed to run various IoT applications such as manufacturing, smart cities, smart homes, smart grid, smart farming, monitoring sensors, actuators, cameras, the Internet of Vehicles (or intelligent transport systems), social media, healthcare, and wireless communication networks. For example, sources of BD may include Internet data (emails, social media, and weblogs), network data, mobile networks or telecoms, machine-to-machine data or the IoT (sensor data), online transactions, medical records, and open data (mostly by governments). It is typically unstructured (such as text, audio, video) or semi-structured (such as emails, tweets, weblogs).

3.4 Wireless BD

Wireless networks play an important role in BD generation and processing. BD in the fifth generation (5G) wireless communication system is complex and challenging. A framework for wireless BD consists of four layers [7]: data, transmission, network, and application layers. In the data layer, wireless BD exhibits spatial-temporal dimensions. It is a high-volume, high-velocity, and high-variety information that requires special information processing tools. It is distinguished by its unique multi-dimensional, personalized, multi-sensory, and real-time features [8]. In the transmission layer, wireless channels are responsible for conveying the information sent from the transmitter to the receiver. Wireless communications show unique advantages over wired communication, such as cost efficiency and flexibility. Wireless network virtualization ensures BD transmission in a wireless network environment. In the network layer, transmission optimization through BD could be implemented. The network layer related wireless is perhaps the most investigated area in wireless BD. Potential applications of wireless big data include smart grids, IoT, mobile computing, and unmanned aerial vehicle (UAV) [9].

3.5 BD Analytics

BD is a problem in search of a solution. By definition, BD is data that are too large for traditional analysis methods and techniques. The process of examining big data is often referred to BDA. Analytics is the application of science to analysis. It is the process of analyzing large data using statistical models, data-mining techniques, and computing technologies.

BDA involves examining large amounts of data to uncover hidden patterns, correlations, actionable insights, trends, and other useful information. It refers to how we can extract, validate, translate, and utilize BD as a new currency. It is an emerging field that is aimed at creating empirical predictions. Data-driven organizations use analytics to guide decisions at all levels. Common BDA techniques include the following [10,11]:

1. *Data mining.* Data mining (DM) is searching for useful information in massive amounts of data. It involves extracting knowledge from large databases and applying the knowledge in making fact-based decisions. DM techniques enable humans to find useful hidden trends and relationships in massive data. DM algorithms allow us to refine data and find their real value. They deal with a large amount of data with great processing speed. Visualization interprets machine

language to humans, while data mining is the mother tongue of machine. Data mining can be predictive or descriptive. Descriptive data mining describes the general characteristics of the data in the database, while the predictive data mining uses the current data to make some prediction. DM techniques include cluster analysis, association rule of learning classification, and regression [12,13].

2. *Web mining.* This is the technique for extracting useful information obtained from the World Wide Web. It is the application of data-mining techniques for discovering patterns from large web repositories. It may be used to learn about customer behavior, evaluate the effectiveness of a specific website, help quantify the success of a marketing strategy, and reveal unknown knowledge about a particular website. Web mining may include content mining, hyperlink structure mining, and usage mining [14,15].

3. *Text analytics.* It is generally accepted that structured data represent only 20% of the information available to an organization, while 80% of all the data are in unstructured form. Also known as text mining or natural language processing, text analytics is the science of turning unstructured text into structured data. Its purpose is deriving high-quality structured data from unstructured (textual) information.

4. *Predictive analytics.* This is the process of predicting future with BDA tools by analyzing current and historical facts. This makes it possible for analysts to make sound decisions based on visualization analysis and data mining. It builds models for forecasting customer behavior and other future developments.

5. *Machine learning.* This is one of the main drivers of the BD revolution because of its ability to learn from data and provide decisions, insights, and trends. Machine learning (ML) is about learning some properties of a data set and applying them to new data. It is the scientific discipline dealing with the ways in which machines learn from experience.

   It refers to the automated detection of meaningful patterns in a given data. It is an incredibly powerful tool that is used in a wide range of compelling application domains because it is applicable to many real-life problems. ML has already shown its capacity to learn and analyze, beating our best champions at complex strategy games, such as *Go* and *Jeopardy!* Common ML algorithms include artificial neural networks, fuzzy logic, and genetic algorithms [16,17].

6. *Deep learning.* Deep learning (DL) refers to a family of approaches that have taken ML to a new level, helping computers make sense out of vast amounts of data. DL algorithms are used to train deep networks with large amounts of data. DL has become a big wave of technology trend for BD and artificial intelligence [18].

7. *Visual analytics.* The main objective of visual analytics is to develop visual and interactive tools and techniques for reasoning and decision-making from large data. The tools and techniques allow decision makers to gain insight into complex problems by to combining their creativity, and background knowledge with the enormous storage and processing capacities of today's computers.

Visual analytics is different from data visualization. Data visualization is the art of placing data in a visual context. The aim of data visualization is to identify trends, patterns, and contexts that usually go unrecognized in text-based data. Visual analytics intends to lean more toward computation and analytical reasoning. It involves critical thinking, sense making, and various analytical methods learned from the intelligence community. Both aim to make data more understandable and more effective. Visual analytics are efficient when working in a geospatial domain and multi-dimensional analysis [19].

8. *Crowdsourcing.* This is the process of getting work done by online community or crowd of people in the form of an open call, the voluntary undertaking of a task. This tool is used more for collecting data than for analyzing it.

9. *Mobile analytics.* Mobile analytics research is emerging in different areas such as mobile sensing apps that are location-aware and activity-sensitive. The ability to collect finegrained, location-specific, context-aware, highly personalized content through these smart devices has opened new opportunities [19a].

Analyzing huge volumes of data improves data-driven decision-making mechanism and the ability to predict future outcomes. BDA helps organizations in significant cost reduction, better decision-making, and creation of new products and services.

3.6 Tools for Analyzing BD

The 5V characteristics of BD will bring software based on traditional approaches to their knee. Once the BD is ready for analysis, we use advanced software programs. These include Hadoop, MapReduce, MongoDB, and NoSQL databases.

- *Hadoop.* This is an open-source software framework for distributed storage of large datasets on computer clusters. It was originally developed in 2006 by Doug Cutting and Mike Cafarella. It provides large amounts of storage for all sorts of data along with the ability to handle virtually limitless concurrent tasks. Hadoop is written in Java and provides Java classes and APIs to access them. It is designed to process large amounts of structured and unstructured data. Although Hadoop is a free and an open-source software, the free version of Hadoop is not easy to use. A number of companies have developed friendlier versions of Hadoop, and Cloudera is the most popular of them all.

- *MapReduce.* This works on Hadoop framework. It is a Google technology for processing massive amounts of data. It is a software framework that enables developers to code programs that can process large amounts of unstructured data. It is a data processing algorithm that involves distributing a task across multiple nodes running a "map" function. It has two components: (1) Map which distributes the input data to several clusters for parallel processing, (2) Reduce which collects all sub-results to provide the final result.

- *NoSQL.* This is an open-source software designed for use in BD application in clustered environments, providing high speed access unstructured or semi-structured data. It provides capabilities to query and retrieve unstructured and semi-structured data. It is available in both an open-source community edition and in a priced enterprise edition.

- *MongoDB.* This is a good resource for managing data that are frequently changing or unstructured. It is a version of NoSQL. It is the modern, start-up approach to relational databases. As with any database, one needs to know how to query it using programming language. It is flexible, highly scalable database designed for web applications. It is often used to store data in mobile apps, product catalogs, and real-time applications.

Other tools include Hive, Cassandra, Spark, Tableau, Talend, and cloud computing. BD computing in clouds is known as "big data clouds."

3.7 Data Quality

Although BD is often characterized in terms of it volume (quantity), the feature of critical importance is the quality of information embodied in the data. Quality of data refers to the meaning of the data, the consistency of the meaning, the human interpretation of results, and the contexts in which the results are used. It is affected by different factors in the process of generation, acquisition, and storage. BD quality varies from one type of BD to another. Data sources may be large and heterogeneous. They may be raw, incomplete with some missing values. All large data sets have errors in them. This is particularly true of data sets collected by human beings. Items may be missing or recorded wrong or may be out-of-range. Data mining can be vulnerable to problems arising from poor-quality data [20].

Quality is an important issue for all data (big or small) because it affects the quality of analysis. For any BD project to succeed, it must depend on high-quality data. The value of data for making decisions may be severally affected by inaccurate data. Quality issue will become a critical factor in Data Warehousing and OLAP methodologies over BD. Quality issues include accuracy, completeness, consistency, precision, relevance, and timeliness. Data are a neutral measure of reality, but people corrupt data and use them to achieve their own ends. Maintaining the quality of data is the challenging task of data analyst.

3.8 Application Domains

Several domains can benefit from the BD phenomena, including IoT, smart power grid, science, healthcare, business, industry, manufacturing, and public agencies.

- *Internet of Things*. The recent emergence of the IoT, the global network of billions of interconnected devices and sensors, has caused an explosion in the volume of data in the form of text, video, images, and even audio. In other words, IoT is one of the biggest sources of BD. Among smart cities constructed based on IoT, BD may come from home, industry, government, etc. Due to the variety of objects involved, this application is endlessly evolving [21].

- *Smart grid*. The smart grid is emerging as a promising technology meant to cope with the energy efficiency problem and integrate renewable energy in the grid. The data produced in smart grids exhibits the 5Vs characteristics of BD. Research and development on smart grids are conducted by utility companies, university communities, and standards bodies. The smart grid can be divided into three major domains: the home area network (HAN), neighborhood area network (NAN), and wide area network (WAN). BD technology is the major factor for the construction of smart power grid [22].

- *Science*. Science has been dealing with large volume of data in research experiments. Scientific research includes collection of data which aim at verifying some scientific hypothesis. For example, a huge number of chemical structures are stored in public and private databases.

- *Healthcare.* Applications of BD are helpful for improving clinical decision-making and care provision. BD is driving the development in biomedical and healthcare informatics because BD has unlimited potential for storing, processing, and analyzing medical data. For example, BD can be leveraged to detect fraud, abuse, and errors in health insurance claims [23]. Healthcare professionals can apply BDA in the same way as other businesses, except that the stakes are higher. Data analytics in healthcare has the potential of transforming the ways healthcare organizations operate, both for business operations and health management. It can improve the care, save patient's life, and lower healthcare cost. The more data are available to physicians, the easier these are to identify trends and identify bottlenecks in patient care. BD is helping to solve healthcare problems in hospitals around the world, leading to reduced waiting times for patients and better quality of care. Today BD allows for early identification of individual patient's illnesses.

- *Nursing.* Nurses are care coordinators and an integral part of the healthcare system. The discipline of nursing is charged with several core missions including protecting, promoting human health, alleviating suffering, and preventing injury and illness. The discipline needs to maximize the benefits of BD to advance the vision of promoting human health and wellbeing. Nursing needs BD, and BD needs nursing. Since BD techniques have the potential to revolutionize health professional education, nursing students should be involved in the effort to increase BD literacy in nursing.

- *Business.* Business or commerce is the major source of BD. Service supply chain such as finance, banking, insurance, tourism, and telecommunications drive BD. In business, BD may be considered as cost-effective techniques for solving business problems whose resource requirements exceed the capabilities of traditional computing environment. BD offers competitive advantage to a company in that it allows the company to outperform its competitors. Although companies may share third party facilities, such as clouds, they do not share data but ensure that their data are protected. Applying BD in business can enhance efficiency and competitiveness in many aspects such as marketing, banking, supply chain, and e-commerce. BD has emerged as a vital means for enhanced business activities, better execution of operations, and improved marketing strategy. Without doubt, BD is making a positive impact on the business sector [24]. BD is viewed as a game changer that is capable of revolutionizing the way many businesses operate and compete.

- *Industry.* In industry, BD may involve controlling technological processes and facilities. BD is being used by industries to enhance manufacturing, fine-tune pricing, improve marketing, and support innovation. It is creating great market opportunities in various industries. Today's organizations are handling increasing amounts and complexities of data. Industry leaders are beginning to revolutionize the decision-making process with BD by dramatically improving productivity and profitability.

- *Manufacturing.* The modern computer-aided manufacturing process is complex. It produces huge amount of data which may need to be stored to allow effective quality control. To meet the growing demand in the market and maintain a competitive edge, manufacturers have started looking for ways in which BD can open a new array of opportunities for an industry. The manufacturing industry can use BD to address existing challenges and simultaneously create new business opportunities. By applying advanced analytical tools to raise their productivity, manufacturers can increase efficiency and enhance product quality [25].

- *Public agencies.* The US government has implemented BD programs through different agencies. A typical public agency such the US Social Security Administration collects, manages, and curates large volumes of data to provide services to US citizens. The agency has made great strides in using BD space to improve administration and delivery of services. This has included: (1) improving its arcane legacy system, (2) developing employee and end-user capability, (3) implementing data management strategies and organizational architecture, (4) managing security and privacy issues, and (5) advocating for increased investment in BDA [26]. BDA in national security and law enforcement have the potential to reap great benefits for the society.

Other applications include education, finance, social science, marketing, retail, gaming, insurance, telecommunications, smart cities, metrology, government, agriculture (smart farming), and surveillance. BD has become ubiquitous across all areas of research allowing for new applications that present their own unique challenges and were not possible earlier.

In addition, social networking and government systems also contain large amount of data. BD applications are gaining momentum as more companies seek to monetize the data and move their business forward.

3.9 Benefits

There are several benefits (potential and realized) of BD and BDA. These include the following [27]:

1. It promises to lead to making better decisions and improves insights and predictions. This can lead to greater operational efficiency, productivity, reduced cost, and risk.
2. It eliminates the biases people have when making decisions based on limited information.
3. Allowing analysis of data to be built into the process that enables automated decision-making.
4. It helps in reducing rates of return, producing high-quality products.
5. Proper application will improve overall profitability of business.
6. It helps social media, public and private agencies to explore behavioral patterns of people.
7. As a company invests in a new technology such as BD, it needs to develop new skills and hire new people who can handle the technology.
8. It can potentially be used in driving economic growth in developing world.

3.10 Challenges

Realizing BD advantages is not an easy task. BD is fraught with a number of challenges that need to be addressed before such real-time BD applications become feasible [28].

- *Complexities.* Processing, storage, and transfer of a large scale of data are a great challenge. BD is a problem for many organizations because current data processing technologies cannot process BD effectively. Different data sources have different data structures, making the integration of such data a non-trivial problem. It is difficult to store and integrate distributed provenance. Some data sources, such as sensor networks, can produce data that are of no interest and it is a challenge to filter out the useless information without discarding useful information.
- *Privacy.* A major risk in BD is leakage of data, which is not good for privacy. The rise of BD raises fundamental challenges in privacy and

data ownership. Privacy concern arises continue from the users who outsource their private data into the cloud storage. Outsourcing data to cloud means that the customer will forfeit physical control over their data. This concern has become serious with the development of BD analytics, which require personal information. But quantifying privacy is very difficult [29].

- *Security.* Concerns are being expressed over the impact that collecting, storing, and processing large amount of data could have on security. This is of major concern due to the variety and heterogeneity of data, increased attack, and grave consequences of breaches. Security of BD is a primary concern for many applications. From security point of view, BD may seriously weaken confidentiality. Security is a concern because of the variety and heterogeneity of BD; there is access to data from multiple and diverse domains. Common security solution involves encrypting the data. It may also demand new legislation including liability and sanctions for infringements [30].
- *Data migration.* A major challenge of BD management is transferring such data for distributed processing and storage. To transfer BD from one geographic location to another is costly. Processing the data located at different geographically distributed centers takes a lot of bandwidth of the data center network [31]. For data integration, it is important to follow standards. Unfortunately, there is a general lack of standardization on how data are stored and processed.
- *Shortage of data scientists.* There is a global shortage of data scientists who use BD analytic tools to capture and analyze BD. In response to a shortage of highly trained big data professionals, a number of universities have created masters programs on BD.

BDA is based on parallelism in which large data are stored and processed in different clusters, which are distributed servers around the world. BD is potentially global, crossing national boundaries. A major challenge is being able to distribute storage and processing according to the regulations and data sensibility [32]. Some of the major obstacles of wireless BD signal include processing and network design with respect to the scale of problem size and the complex problem structures. In spite of these challenges, BD is not only promising but also inevitable in view of the persistent data volume explosion.

3.11 Conclusion

The massive growth in mobile networks, cloud computing, and social networking has ushered in the so-called BD revolution with substantial opportunities. Large data volumes are generated on daily basis at unprecedented rate from heterogeneous sources such as social networks, health, government, marketing, and financial institutions. *BD* refers to data that would typically be too expensive to collect, store, manage, share, and analyze using traditional database systems.

BD also refers the technologies that make processing and analyzing it possible. It has attracted a growing attention from government, industry, and academia. It has emerged as a full-fledged field. It is a pretty new research area for both public and private organizations. Universities are beginning to cover BD in their curriculum [33].

BD should not be ignored by organizations and engineers since it is here to stay and will have an increasing impact in all sectors of our society. It is poised to change our decision-making culture and transform how we live, work, and think. It will become a key basis for innovation, growth, competition, and productivity. Scholars can stay up to date on issues related to big data by consulting some recent books [34–37], several books on BD available in Amazon.com, and *IEEE Transactions on Big Data, Journal of Big Data*, and *Big Data & Society*, the three journals exclusively devoted to BD.

References

1. N. Jukic et al., "Augmenting data warehouses with big data," *Information Systems Management*, vol. 32, 2015, pp. 200–209.
2. M. N. O. Sadiku, M. Tembely, and S. M. Musa, "Big data: an introduction for engineers," *Journal of Scientific and Engineering Research*, vol. 3, no. 2, 2016, pp. 106–108.
3. H. Fang et al., "A survey of big data research," *IEEE Network*, vol. 29, no. 5, September/October 2015, pp. 6–9.
4. D. Lancy, "3D data management: controlling data volume, velocity, and variety," *META Group*, vol. 6, no. 70, 2001, p. 1.
5. P. Del Vecchio et al., "Big data for open innovation in SMEs and large corporations: trends, opportunities, and challenges," *Creativity and Innovation Management*, July 2017, pp. 1–17.
6. J. Moorthy et al., "Big data: prospects and challenges," *The Journal for Decision Makers*, vol. 40, no. 1, 2015, pp. 74–96.
6a. H. Asri et al., "Big data in healthcare: challenges and opportunities," *Proceeding of International Conference in Cloud Technologies and Applications*, June 2015.
7. L. J. Qian, J. K. Zhu, and S. H. Zhang, "Survey of wireless big data," *Journal of Communications and Information Networks*, vol. 2, no. 1, 2017, pp. 1–18.
8. X. Cheng et al., "Mobile big data: the fuel for data-driven wireless," *IEEE Internet of Things Journal*, vol. 4, no. 5, 2017, pp. 1489–1516.
9. "Big data analytics," http://searchbusinessanalytics.techtarget.com/definition/big-data-analytics.

10. M. N. O. Sadiku, S. R. Nelatury, and S. M. Musa, "Wireless big data," *Journal of Scientific and Engineering Research*, vol. 4, no. 9, 2017, pp. 107–110.

11. I. Yaqoob et al., "Big data: from beginning to future," *International Journal of Information Management*, vol. 36, 2016, pp. 1231–1247.

12. M. N. O. Sadiku, A. E. Shadare, and S. M. Musa, "Data mining: a brief Introduction," *European Scientific Journal*, vol. 11, no. 21, July 2015, pp. 509–513.

13. M. N. O. Sadiku, S. M. Musa, and O. M. Musa, "Data mining in the chemical industry," *International Journal of Trend in Research and Development*, vol. 4, no. 6, November/December 2017, pp. 295–296.

14. R. Gupta, "Journey from data mining to web mining to big data," *International Journal of Computer Trends and Technology*, vol. 10, no. 1, April 2014, pp. 18–20.

15. P. B. Mohata, "Web data mining techniques and implementation for handling big data," *International Journal of Computer Science and Mobile Computing*, vol. 4, no. 4, April 2015, pp. 330–334.

16. M. N. O. Sadiku, S. M. Musa, and O. S. Musa, "Machine learning," *International Research Journal of Advanced Engineering and Science*, vol. 2, no. 4, 2017, pp. 79–81.

17. A. L'Heureux et al., "Machine learning with big data: challenges and approaches," *IEEE Access*, vol. 5, 2017, pp. 7776–7797.

18. M. N. O. Sadiku, M. Tembely, and S. M. Musa, "Deep learning," *International Research Journal of Advanced Engineering and Science*, vol. 2, no. 1, 2017, pp. 77–78.

19. A. Moran et al., "Improving big data visual analytics with interactive virtual reality," *Proceedings of IEEE High Performance Extreme Computing Conference*, September 2015.

19a. H. Chen, R. H. L. Chian, and V. C. Storey, "Business intelligence and analytics: from big data to big impact," *MIS Quarterly*, vol. 36, no. 4, December 2012, pp. 1165–1188.

20. A. Ganapathi and Y. Chen, "Data quality: experiences and lessons from operationalizing big data," *Proceedings of IEEE International Conference on Big Data*, 2016, pp. 1595–1602.

21. E. Ahmed et al., "The role of big data analytics in the Internet of Things," *Computer Networks*, vol. 129, 2017, pp. 459–471.

22. C. Tu et al., "Big data issues in smart grid – a review," *Renewable and Sustainable Energy Reviews*, vol. 79, 2017, pp. 1099–1107.

23. R. Schroeder, "Big data business models: challenges and opportunities," *Cogent Social Sciences*, vol. 2, 2016, pp. 1–15.

24. M. Viceconti, P. Hunter, and R. Hose, "Big data, big knowledge: big data for personalized healthcare," *IEEE Journal of Medical and Health Informatics*, vol. 19, no. 4, July 2015, pp. 1209–1215.

25. M. N. O. Sadiku, S. M. Musa, and O. S. Musa, "Big data in the chemical industry," *International Journal of Advances in Scientific Research and Engineering*, vol. 3, no. 10, November 2017, pp. 20–23.

26. R. Krishnamurthy and K. C. Desouza, "Big data analytics: the case of the social security administration," *Information Polity*, vol. 19, no. 3–4, 2014, pp. 165–178.

27. N. Garg, S. Singla, and S. Jangra, "Challenges and techniques for testing of big data," *Procedia Computer Science*, vol. 85, 2016, pp. 940–948.

28. H. Barham, "Achieving competitive advantage through big data: a literature review," *Proceedings of Portland International Conference on Management of Engineering and Technology (PICMET)*, Portland, Oregon, July 2017.

29. A. Mehmood et al., "Protection of big data privacy," *IEEE Access*, vol. 4, 2016, pp. 1821–1834.
30. Y. Gahi, M. Guennoun, and H. T. Mouftah, "Big data analytics: Security and privacy challenges," *Proceedings of IEEE Symposium on Computers and Communication*, 2016.
31. P. Teli, M. J. Thomas, and K. Chandrasekaran, "Big data migration between data centers in online cloud environment," *Procedia Technology*, vol. 24, 2016, pp. 1558–1565.
32. T. van den Brock and A. F. van Veenstra, "Governance of big data collaborations: how to balance regulatory compliance and disruptive innovation," *Technology Forecasting and Social Change*, 2017.
33. B. Due et al., "Introducing big data topics: a multicourse experience report from Norway," *Proceedings of the 3rd International Conference on Technological Ecosystems for Enhancing Multiculturality*, Porto, Portugal, 2015, pp. 565–569.
34. R. Kitchin, *The Data Revolution: Big Data, Open Data, Data Infrastructures and Their Consequences*. London, UK: Sage Publications, 2014.
35. N. Marz and J. Warren, *Big Data: Principles and Best Practices of Scalable Real-time Data Systems*. Shelter Island, NY: Manning Publications, 2015.
36. A. Bahga and V. Madisetti, *Big Data Science & Analytics: A Hands-On Approach*. Johns Creek, GA: Self-published by the authors, 2016.
37. J. Hurwitz et al., *Big Data for Dummies*. Hoboken, NJ: John Wiley & Sons, 2013.

4

Cloud Computing

Bitterness is believing God got it wrong, worry is believing God will get it right, and unforgiveness is believing that you are right even after God says you are wrong.

Randle Lowrance

4.1 Introduction

Cloud computing (CC), or simply "the cloud," is a new wave in the field of information technology. Some see it as an emerging field in computer science. Computation, which used to be confined to one location, is now centralized in vast shared facilities. CC impacts how we share, store, retrieve, and process information privately and corporately [1].

CC is an emerging computing paradigm for delivering computing services (such as servers, storage, databases, networking, software, analytics, and more) over the "the cloud" or Internet with pay-as-you-go pricing. Hence, "cloud computing" is also called "Internet computing." The term "cloud computing" was introduced in 1961 when computer scientist John McCarthy predicted that computing would become a public utility. CC was first commercialized in 2006 by Amazon's Elastic Computer Cloud (EC2). The word "cloud" is a metaphor for describing web as a space where computing has been preinstalled and exist as a service. It originated from the habit of drawing the Internet as a fluffy cloud in network diagrams. The cloud enables you to access your information from anywhere at any time. Figure 4.1 shows a conceptual diagram of CC [2].

CC is a means of pooling and sharing hardware and software resources on a massive scale. Users and businesses can access applications from anywhere in the world at any time. Companies offering these computing services are called cloud providers and typically charge for CC services based on usage. Their aim is making computing a utility such as water, gas, electricity, and telephone services.

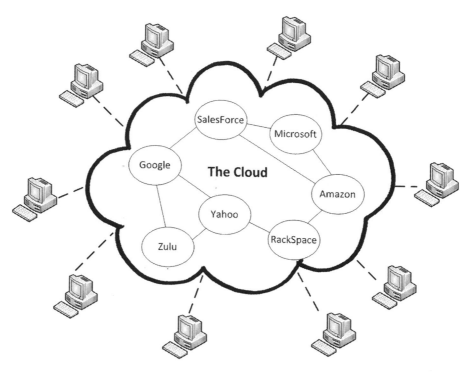

FIGURE 4.1
Cloud computing [2].

The main objective of CC is to make a better use of distributed resources and solve large-scale computation problems. For example, CC can focus the power of thousands of computers on one problem enabling researchers to do their work faster than ever. It may be multiple users collaborating on documents in real time across the Internet. It may also be a web-based software which completely eliminates the need to install the software on every user's computer. Thus, CC may be regarded as a distributed system that offers computing services via a computer communication network, usually the Internet (TCP/IP). Resources in the cloud are transparent to the users and the users need not know the exact location(s) of the resources. They can be shared among a large number of users, who should be able to access applications and data from anytime anywhere (e.g., different work locations, while traveling, etc.).

This chapter provides a comprehensive introduction to CC. It begins by looking at the essential characteristics and technological enablers of CC. It covers different cloud services, deployment models, mobile CC, and cloud service provides. It examines the benefits and challenges of CC. The last section provides some conclusions.

4.2 Key Characteristics

The CC architecture consists of a massive network of interconnected servers, often with a user-friendly front–end interface, which allows users to select services. CC allows access to the collection of computing resources via the Internet. Software and associated resources are hosted in the cloud and can be assessed by the user's computing devices such as desktop, laptop, smartphone, and tablet.

The National Institute of Standards and Technology (NIST) defines general characteristics of CC as follows [3]:

- *Automated self-service set up*. Users can run and configure their own computing resources without human help.
- *Broad network accessibility*. Computing resources are available through the Internet on various devices.
- *Pool resources*. Users do not have their own dedicated hardware. Multiple users can employ the same hardware and resources.
- *Scalability*. This refers to the ability of a system to perform well under different load sizes. User can easily expand the amount of power available to them. Applications that run within the cloud can be configured to scale automatically. CC is adoptable for all sizes of businesses.
- *Metered services*. Use of cloud resources is monitored and users are charged for the amount of resources they use. The user pays only for the storage or bandwidth consumed, and not for potential bandwidth.
- *Rapid elasticity*. A user can gain resources from the cloud and rapidly discharge those resources when they are no longer needed.

4.3 Technological Enablers

CC is essentially a combination of many preexisting technologies. The key technological enablers for CC include the following: virtualization technologies, grid computing, distributed computing, multitenancy, and elasticity [4–6].

1. *Virtualization*. The key feature of CC is the idea of virtualization, which enables an operating system to run on several hardware deployments. It is one of the technologies that enable elasticity. Virtualization

decouples the software from the hardware. It is the technology that uses a physical resource such as a server and divides it into virtual resources known as virtual machines. There are three forms of virtualization: server virtualization, storage virtualization, and network virtualization. Server virtualization increases resource sharing by masking server resources like processors, operating systems, RAM, etc., for server users. Storage virtualization is pooling of physical storage from storage devices into a single storage device which is managed centrally. Network visualization is a means of combining resources in a network by splitting up the available bandwidth into independent channels. Several cloud infrastructures are built with virtual servers and virtual machines. Within a virtualized environment, some networking functionalities can stay on a physical server. CC lowers cost of operation by employing virtual machines, which are managed by the hypervisor. The hypervisor allocates resources to every virtual machine. It therefore becomes the single point of failure.

2. *Grid computing.* Grid computing refers to a distributed architecture of a large number of computers connected to solve a complex problem. It allows the computers on the network to work on a task together, behaving like a supercomputer. Grid technologies leverage the computational power of the available computers by managing them in the grid infrastructure. Grid or distributed computing is a special type of parallel computing that relies on complete computers connected to a computer network. The grid may be small or large. As distributed computing, grid computing was motivated by the electrical power grid. CC is regarded as an evolution of grid computing. It provides scalable computing resources on payment basis [7].

3. *Distributed computing.* CC is a distributed computing paradigm, which allows for highly elastic resource pool. Clouds are built to share computing, share memory, share storage, and other resources. Load balancing establishes an algorithm for assigning tasks to the cloud nodes. All computers in distributed systems are quite independent and do not share resources. Each computer has its own responsibility.

4. *Multitenancy.* This involves sharing a single set of infrastructure across several customers and stakeholders. Virtualization helps to deliver infrastructure multitenant capability. Multitenancy implies that multiple tenants share computational resources, storage, database, services, computing, memory, and other resources with other tenants. This sharing violates the confidentiality of tenants. For secure multitenancy, there should be a degree of isolation among tenant data.

5. *Elasticity.* This refers to the ability to automatically scale up and handle high volumes of traffic or scale down and use less resources when needed, maximizing the use of resources. It implies that users can grow or shrink infrastructure resources dynamically based on

the current demand. When automated, elasticity allows a cloud provider to continuously monitor a customer's infrastructure and scale it on-demand.

4.4 Cloud Services

CC is not a single produce or piece of technology. Rather, it is a system, primarily providing three different services. The services provided by CC are shown in Figure 4.2 and explained as follows.

- *Infrastructure-as-a-Service* (IaaS). This is the simplest of CC offerings. It involves the delivery of huge computing resources such as the capacity of storage, processing, operating systems, servers, computing power, firewalls, bandwidth, and network which form the underlying cloud infrastructure. It allows users to rent any form of hardware and software and remotely access computing resources on a pay-per-use basis. The major advantages of IaaS are pay per use, security, and reliability. IaaS is also known as Hardware-as-a-Service (HaaS). An example of IaaS is the Amazon EC2. IBM and Amazon use IaaS.
- *Platform-as-a-Service* (PaaS). This supports the development of web applications quickly and easily. The customer can build his own applications, which run on the cloud provider's infrastructure. It has emerged due to the suboptimal nature of IaaS for CC and the

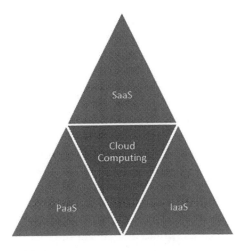

FIGURE 4.2
Cloud computing services.

creation of web applications. Many big companies seek to dominate the platform of CC, as Microsoft dominated personal computer (PC). Examples of PaaS are Google App Engine and Microsoft Azure.

- *Software-as-a-Service* (SaaS). This is also known as Software-as-a-Server or Software-on-demand. This provides a service (software applications over the Internet) that is directly consumable by the end-user. It is a software deployed over the Internet. This is a pay-as-you-go service. It seeks to replace the applications running on PC. Google, Twitter, and Facebook are typical examples of SaaS. Gmail, AOL, Yahoo, and Skype all use SaaS.

These are often called the CC stack since they build on top of one another.

The service models are useful in categorizing not only CC, but specific vendor offerings, products, and services. CC delivers services based on demand and the consumers only use the services they need. As CC becomes mature, several service types are being introduced and overlaid on these architectures. These include HaaS, application-as-a-service (AaaS), network-as-a-service (NaaS), data-storage-as-a-service (dSaaS), IT-as-a-service (ITaaS), functions-as-a-service (FaaS), cooperation-as-a-service (CaaS), and municipality-as-a-service (MaaS).

4.5 Deployment Models

CC is the environment which enables users and businesses to employ application on the Internet such as retrieving, storing, and protecting their personal file or data. Application and databases are moved to the centralized data centers, known as clouds. There are three deployment models of cloud. Figure 4.3 shows their relationship [8].

- *Public cloud.* This describes CC in the traditional mainstream sense. It is characterized by public availability of the cloud services. It allows users to access the cloud through interfaces using web browsers. It is when the services are provided over a network that is open for the public. It is a cost-effective, elastic means of deploying a solution. It is typically based on a pay-per-use model. It is less secure than the other cloud models. Typical examples of public cloud are Google and Facebook. Public cloud service providers operate the infrastructure at their data center and provide access through the Internet.

- *Hybrid cloud.* This is the cloud environment that offers the benefits of multiple deployment models. It provides solutions through a mix of both private and public clouds. Its main strength is that it provides

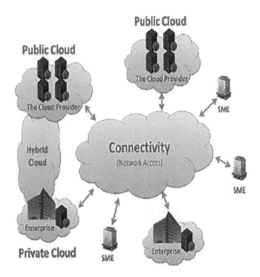

FIGURE 4.3
Deployment model types [8].

better scale and convenience of a public cloud as well as the control and reliability of private CC. Its adoption depends on several factors such as data security, compliance requirements, and the level of control needed over data.

- *Private cloud.* This describes offering that makes cloud services available for a designated single, private organization. It is created within an organization's data center. It is different from public cloud in that all the resources and applications are managed by the organization. It is more secure than the public cloud because only the organization stakeholders are allowed to access the private cloud. When appropriately constructed, it can improve business. Eucalyptus was first to deploy private clouds.

Apart from these, other CC deployment models include community CC, distributed cloud, shared private CC, dedicated private CC, multicloud, and dynamic private clouds.

4.6 Mobile Cloud Computing

Mobile devices are used widely nowadays. The number of mobile users is increasing due to the constant improvement of user-friendly mobile devices. Mobile CC is quickly gaining popularity among mobile device users as it is able to offer CC capabilities in a mobile environment. A massive uptake of CC in the developing world is likely to be through mobile devices. Mobile devices have become indispensable part of information and communication technology. They are currently limited by battery life, storage, bandwidth, and power. A mobile application involves a software designed to run on a mobile device. The new generation of mobile applications will rely on cloud infrastructures.

Mobile CC combines CC, mobile computing, and wireless networks to make computational resources available to mobile users and network operators. It integrates CC with mobile devices and combines the advantages of both mobile computing and CC [9]. Although mobile users benefit from storing their data on the cloud, they should be concerned about data integrity and authentication. The integration of CC, mobile devices, and wireless mobile networks raises issues such as trust, security, privacy, and limited resources in mobile devices. Applications of mobile computing include mobile healthcare, mobile commerce, mobile learning, mobile gaming, and mobile government [10].

4.7 Cloud Service Providers

The major vendors of CC include IBM, Microsoft, Cisco, Oracle, Sun, and Siemens. In a CC environment, the role of service provider is twofold: infrastructure providers who manage cloud platforms and lease resources, and service providers who rent resources from infrastructure providers to serve the end users. Several companies including IBM, Microsoft, Google, Verizon, Amazon, Rackspace, Eucalyptus, and Netflix have started to offer diverse cloud services to their customers. The focus is on providing IaaS, PaaS, and SaaS based offerings. The top cloud service providers include the following [11]:

1. *Microsoft cloud.* There are many cloud services to the name of Microsoft, including OneDrive, Office 365, and Azure. Window Azure is development, hosting, and management environment that facilitates on-demand computing capability. It is hosted in Microsoft data center and provides operating system. Microsoft also offers a set of web-based apps, Office Online, which are Internet versions of Word, Excel, PowerPoint, and OneNote.

2. *IBM cloud.* IBM has two main offerings: IaaS and SaaS (cloud software). In 2011, IBM announced the IBM SmartCloud framework to support Smarter Planet.

3. *Amazon cloud.* Amazon offers EC2 and Amazon Web Services (AWS). EC2 basically offers a PaaS platform for developers, a place to develop

their apps. Its offerings of cloud include a web service and computing capacity. AWS is 100% public and includes a pay-as-you-go.

4. *Google cloud*. The main ones are Google App Engine and Google Compute. Google App Engine enables developers to design, develop, and deploy Java and Python-based applications.

These are the dominant CC products. Others include Force.com from Salesforce, Dropbox, iCloud (Apple) used globally, Ucloud (KT) emerging in Korea, Mobile office (KT), and Mobile share (AT&T). Charges for cloud services are based on three key items: storage, bandwidth, and compute. Storage is the amount of data (in GB) stored over a monthly period. Bandwidth is calculated at the amount of data (in GB) transferred in and out of platform service through batch processing and transaction. It can grow for data-intensive application. Compute is the time (in hours) needed to run an application.

4.8 Application Domains

CC is an emerging technology to the scientific community and is used in variety of disciplines. The percentages of cloud usage in different industrial sectors and services are shown in Figure 4.4 [12]. CC is used in education, large enterprises, small and medium businesses, manufacturing, industrial automation, life sciences, pharmacy, e-government, and medicine.

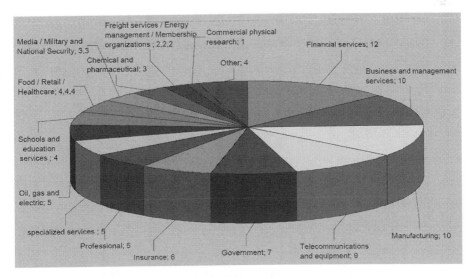

FIGURE 4.4
Percentages for cloud usage in different industrial sectors and services [12].

- *Manufacturing.* Manufacturing is the production of products using raw materials, labors, machines, and tools. Major production facilities of most companies run 24/7. Manufacturing industry has always been a pillar industry of developed economies. It has been a strong impetus to the process of industrialization and modernization. To survive in an increasing competitive pressure, globalization, and rapid technology development, modern manufacturing requires a flexible and dynamic provisioning and management, using manufacturing resources available on demand over computer networks. Technology has always been the main driver of transformation of manufacturing.

  The CC has many benefits to apply their technique to the manufacturing system such as efficiency, flexibility, control, and management for the system [13]. Cloud manufacturing (CM) is emerging as a new manufacturing paradigm which applies well-known basic concepts from CC to manufacturing processes and deliver shared, ubiquitous, on-demand manufacturing services. It is an innovative, web-based manufacturing model. It is promising to transform today's manufacturing industry from production-oriented to service-oriented, highly collaborative manufacturing of the future.

  State-of-the art technologies that enable CM include CC, pay-as-you-go/use utility computing, virtualization, social media, Internet of Things (IoT), industrial Internet, big data, wearable technologies, and service-oriented architecture (SOA). Because IoT is known for ubiquitous computing (using embedded sensors and actuators) and pervasive sensing technologies (such as RFID tags), it is capable of automating manufacturing processes by connecting humans, machines, and manufacturing processes.

- *Pharmaceutical companies.* These companies find CC to be very attractive since it offers them the promise of efficient and cost-effective data analysis and processing as well as multi-layered security. Another factor in favor of CC is its low cost, pay-as-you-grow business model. Early adopters of CC such as Pfizer, Johnson & Johnson, and Eli Lilly all used AWS and Amazon EC2. They were able to perform R&D using the cloud, and process proteomics, bioinformatics, and statistics [14].

- *Oil & gas.* CC can be beneficial to the oil and gas industries in many ways: easy viewing of geological survey data of the entire oil field, tracking any fluctuations in the data and auto-correcting them, expand the business to a larger scale, and choose the best available and appropriate technology that will be useful for the explorers. The oil and gas companies around the globe are embracing the benefits provided by the private clouds since they are too great and numerous to ignore [15].

- *Industrial automation.* Process automation plants can adopt CC to solve business challenges and utilize the rapid spread of Internet-based service models. Applications suitable with CC technology for industrial automation include condition based maintenance, predictive maintenance, and asset management [16].

- *E-Government.* The promise of cloud services meets the interest of several governments worldwide. These governments are conscious of how taxpayer money is spent and are interested in finding ways to do more with less. The government is often the largest IT user in any country. CC can enhance how governments operate and deliver services. The key components of open government are citizen engagement, open data, collaboration, and innovation. Each of these components can utilize the cloud. Some developing nations face tough challenges such as corruption, poor public management, lack of transparency, and insufficient IT skill. CC has the potential to revolutionize e-government and provide a basis to address some the challenges [17–19].

- *Enterprises.* Enterprises always want secure, reliable, and affordable IT infrastructures. To fulfill these requirements, CC is attracting the interest of enterprises, small and large, around the globe and they have started to shift more core business functions onto the cloud platforms. Although adopting CC in enterprise environments is not an easy task, CC saves cost and time for businesses. It provides business agility and makes it easier for enterprises to scale their services according to customer demand [20]. It allows companies to outsource their entire IT process so that they can concentrate more on their core business. Right now, cloud services are available at reasonable prices and accessible for all scales of businesses.

CC is also used in the military, smart grid, entertainment, and geographic information systems (GIS). The adoption of CC in academic institutions is still in its early stages.

4.9 Benefits

CC has the potential to bring a number of benefits to those who adopt it. Top benefits of CC include cost efficiency, speed and agility, productivity, and global scalability.

1. *Cost efficiency.* There is the greatest benefit of adopting CC. CC eliminates or minimizes the capital expense of investing heavily on buying and running hardware and software on-site datacenters and servers. In other words, there is no up-front investment. By

using CC, you can achieve a lower variable cost than you can get on your own. CC lets you stop spending money on running and maintaining data centers and servers. It requires fewer IT staff.

2. *Speed and agility.* CC brings agility, speed, instant availability of resources, and tools that enable collaboration and innovation. In a CC environment, new IT resources are only ever a click away. This results in a dramatic increase in agility for the organization.

3. *Productivity.* CC increases productivity when multiple users can work on the same data simultaneously, rather than waiting for it to be saved and emailed. It enables your employees to work from anywhere anytime as long as they have Internet access. It increases flexibility for the employees.

4. *Ease accessibility.* Data and apps are accessible from any connected computer anywhere, 24/7. User can access the environment of cloud not only from the system but through other several options, which include tablets, iPads, and mobile phones. It not only increases efficiency but enhances the services provided to the consumers. Users can check their email on any computer and even store files using services such as Dropbox and Google Drive. Users find that when they upload their photos, documents, and videos to the cloud and then retrieve them at their convenience, it saves storage space on their desk tops or laptops.

5. *Global scalability.* Scalability is a built-in feature for cloud deployments. With CC, one can go from small to large quickly. CC has the ability to scale elastically, meaning delivering the right amount of IT resources when it is needed, and from the right geographic location.

If used properly, CC is a technology with great opportunity for businesses of all sizes. An industry needs the cloud for the following reasons [21]: (1) Mobile workforce: empowering employees to sift real time data and make decisions on the fly, (2) Minimize disruptions: with the right sort of cloud setup problems can be anticipated and solved quickly, (3) Collaboration: with the right technology, collaboration—as well as transparency and accountability—are easily managed, (4) Innovation: product innovation and process innovation are powerful weapons to survive or thrive in such an environment, (5) Lower cost: No hardware procurement, maintenance, or staff is needed to maintain the systems. CC services are offered on a pay-as-you-go basis.

Other benefits of CC are low cost, flexible resources, speed to market, on-demand self-service, ubiquitous network access, location-independent resource pooling, transference of risk, easy management, cost reduction, green computing, cost saving, flexibility, ease of access, quality of service, rapid scalability, resiliency, and reliability. Although CC offers several benefits, many major players might hold back until some of its challenges are addressed.

4.10 Challenges

As is the case with any new disruptive innovation, there are challenges associated with CC. Those challenges include security, privacy, cost uncertainty, latency, reliability, and regulations.

1. *Security.* This is the greatest challenge for CC technology. Information security is about keeping data secret. CC systems involve networks that are prone to cybersecurity issues such as maintaining data confidentiality, data integrity, and system availability. Some see security as a detractor from using the cloud, especially when it comes to sensitive medical records and financial information. Handing over confidential information to another company makes some feel unsecure. Since users are exposed on the public Internet, they are more vulnerable target for malicious users and hackers. Vendors, researchers, and security professionals are working on security issues associated with CC [22,23].

2. *Privacy.* CC poses privacy concerns because the service provider can access the data that is in the cloud at any time. As data is accessed from any location, user's privacy can be compromised. Users may not know where their data is being stored and who will have the privilege to access it. Some users worry that their personal information can be used for commercial purposes by the service provider [24].

3. *Cost uncertainty.* There is no standard cloud serves pricing. Cloud providers charge differently for data storage, memory, CPU, bandwidth, etc. This creates uncertainty and makes it hard to estimate the financial benefits of CC.

4. *Latency.* For some applications, latency is another issue with CC. This is the delay from request for data to its ultimate delivery. Even a high-bandwidth satellite connection can lead to poor quality performance due to high latency. The quality of service (QoS) of CC still poses significant challenges.

5. *Regulations.* Computing clouds have grown to include users across different nations, often moving personal information across multiple jurisdictions. Some CC providers allocate their data centers in several nations. For example, Google's cloud servers are at different locations in Americas, Asia, and Europe. CC service providers run against international law involving data protection and privacy regulations [25]. Regulation ensures that dominant players do not abuse their position. Compliance regulations may be location-related, security-related, privacy-related, etc. Some government regulations do not allow information about their citizens to

be on a cloud or data center outside their jurisdiction. This is a big challenge for cloud providers. Besides, there is the problem of lack of standards [26].

Another major challenge of moving applications to the cloud is the requirement to master multiple languages and operating environments [27]. Other challenges include loss of technical/organization control, data portability, and software compatibility. Organizations such as Cloud Security Alliance (www.cloudsecurityalliance.org) and Trusted Computing Group (https://trustedcomputinggroup.org) are addressing these concerns and developing standards and specifications for CC. Although the concerns or challenges are slowing down the adoption of CC, CC has rocked the business world. The pros outweighs the cons of CC.

4.11 Future Trends

We are almost at the beginning of the cloud era; it is hard to predict the impact of CC on our society. Even education is moving to the cloud. This allows every student to have access to all of the resources necessary for the class. Although cloud-like infrastructures are here to stay, CC is not fully mature and needs to be explored. The scientific community sees CC as being part of the solution to cope with the burgeoning data volumes.

CC is enjoying tremendous interest in almost all areas of computational use. It has become popular among users and businesses around the globe. It is just emerging and there is room for improvement. Here we consider some sample technologies that will determine the future of CC.

The CC infrastructures are evolving and requiring new computing models to meet the demands of large-scale applications. CC is emerging to accommodate more innovative mobile applications. Volunteer CC (also known as peer-to-peer cloud or ad-hoc cloud) involves groups of people who volunteer their computing resources because they believe in the goals of a given cloud project [28,29].

The Cloud of Things (CoT) has emerged from the integration of CC and IoT. CC is more matured than IoT and can support IoT services and applications. Some new CoT concepts are derived from IoT. Such concepts include Sensing-as-a-Service, Data-as-a-Service, Big Data Analytics-as-a-Service [30].

Cloud computing in VANETs (CC-V) is the next paradigm shift for CC to offer feasible, smart solutions for traffic issues. Vehicular CC is the realization of autonomous cloud among vehicles to share their resources [31].

Software-defined networking (SDN) has emerged as the future networking paradigm. SDN addresses the failure of the traditional networks to support the dynamic, scalable computing and storage needs of today's applications. Through SDN orchestration and CC, network resources can be integrated to provide end-to-end network transport services [32].

4.12 Conclusion

CC has emerged as a key technology for sharing resources over the Internet. It is a disruptive global technology that is massively changing how computing is done. It delivers infrastructure, platform, and software to consumers on a pay-as-you-go manner. As the next generation of on-demand services via the Internet, it has made a significant impact on individuals, health care, education, and e-commerce. It is becoming pervasive and omnipresent in our daily lives; many have used it without being aware of it [33]. It has made its mark and will continue to be popular. Startups, small and medium businesses are using CC to great advantage.

CC is becoming increasingly popular in education at all levels. It has several new opportunities for education. Some high schools and universities have started offering courses on CC. CC offers scalability and flexibility for students, faculties, and staff to access databases, storage, and other university applications anytime anywhere [34–38].

More information about CC can be found in current books [39–45], other books available on Amazon.com, and four international journals devoted to it: *IEEE Transactions on Cloud Computing, International Journal of Cloud Applications and Computing, International Journal of Cloud Computing and Services Science,* and *i-manager's Journal on Cloud Computing.*

References

1. Y. Jadeja and K. Modi, "Cloud computing – concepts, architecture and challenges," *Proceedings of 2012 International Conference on Computing, Electronics and Electrical Technologies* (ICCET), 2012, pp. 877–880.
2. M. N. O. Sadiku, S. M. Musa, and O. D. Momoh, "Cloud computing: opportunities and challenges," *IEEE Potentials,* vol. 33, no. 1, January/February 2014, pp. 34–36.
3. P. Mell and T. Grance, *The NIST Definition of Cloud Computing.* Gaithersburg, MD: National Institute of Standards and Technology, September 2011.
4. S. Azodolmolky, P. Wieder, and R. Yahyapour, "Cloud computing networking: challenges and opportunities for innovations," *IEEE Communications Magazine,* vol. 51, no. 7, July 2013, pp. 54–62.

5. K. E. Kushida, J. Murray, and J. Zysman, "Cloud computing: from scarcity to abundance," *Journal of Industry, Trade and Competitiveness*, vol. 15, February 2015, pp. 5–19.
6. S. M. Hahemi and A. Hanani, "Cloud computing: use cases and various applications," *Journal of Advanced Computer Science and Technology*, vol. 3, no. 2, 2014, pp. 160–168.
7. M. N. O. Sadiku, A. E. Shadare, and S. M. Musa, "Grid Computing," *International Journal of Advanced Research in Computer Science and Software Engineering*, vol. 7, no. 6, June 2017, pp. 5–6.
8. B. Grobauer, T. Walloschek, and E. Stocker, "Understanding cloud computing vulnerabilities," *IEEE Security and Privacy*, vol. 9, no. 2, March/April 2011, pp. 50–57.
9. H. Kaur and N. Bhardwaj, "A review on security issues in cloud computing," *International Journal of Advanced Research in Computer Science*, vol. 6, no. 2, March/April, 2015, pp. 178–182.
10. D. R. R. Joan, "An overture of mobile cloud computing and its applications," *i-manager's Journal on Cloud Computing*, vol. 1, no. 4, August–October 2014, pp. 11–18.
11. "Best cloud computing services," www.thewindowsclub.com/cloud- computing-services.
12. T. Ercan, "Effective use of cloud computing in educational institutions," *Procedia Social and Behavioral Sciences*, vol. 2, 2010, pp. 938–942.
13. H. Y. Jeong and B. W. Hong, "The future manufacturing system and cloud computing," *Applied Mechanics and Materials*, vols. 313–314, 2013, pp. 1357–1361.
14. T. Sommer, "Cloud computing in emerging biotech and pharmaceutical companies," *Communications of the IIMA*, 2013, vol. 13, no. 3, 2013, pp. 37–53.
15. M. A. H. Khan, S. Jailkhani, and B. G. P. Kumar, "Applications of cloud computing in remote oil and gas operations," *Proceedings of 2012 International Conference on Cloud Computing, Technologies, Applications & Management*, 2012, pp. 52–55.
16. D. S. Latha and K. Javaprakash, "Cloud computing technology and its applications in process industry automation," *Chemical Industry Digest*, January 2017.
17. F. Mohammed et al., "Cloud computing fitness for e-government implementation: importance-performance analysis," *IEEE Access*, 2017.
18. F. Mohammed et al., "Cloud computing adoption model for e-government implementation," *Information Development*, vol. 33, no. 3, 2017, pp. 303–323.
19. K. Irion, "Government cloud computing and national data sovereignty," *Policy and Internet*, vol. 4, no. 3–4, 2012, pp. 40–71.
20. M. G. Avram, "Advantages and challenges of adopting cloud computing from an enterprise perspective," *Procedia Technology*, vol. 12, 2014, pp. 529–534.
21. S. Guertzgen, "SAPPOV: why chemical companies need the cloud," December 2013, www.digitalistmag.com/industries/chemicals/2013/12/17/sappov-chemical-companies-need-cloud-01240961.
22. M. A. Khan, "A survey of security issues for cloud computing," *Journal of Network and Computer Applications*, vol. 71, 2016, pp. 11–19.
23. W. O. Victor et al., "A survey on mobile cloud computing with embedded security considerations," *International Journal of Cloud Computing and Services Science*, vol. 3, no. 1, February 2014, pp. 53–66.

24. Y. Sun et al., "Data security and privacy in cloud computing," *International Journal of Distributed Sensor Networks*, vol. 10, no. 7, July 2014, pp. 190–903.
25. I. Verma, "Cloud computing: a study of benefits and challenges," *International Journal of Advanced Studies in Computer Science and Engineering*, vol. 3, no. 7, 2014, pp. 14–17.
26. S. Song, "Competition law and interoperability in cloud computing," *Computer Law & Security Review*, vol. 33, 2017, pp. 629–671.
27. B. Hayes, "Cloud computing," *Communications of ACM*, vol. 51, no. 7, July 2008, pp. 9–11.
28. A. Marosi, J. Kovacs, and P. Kacsuk, "Towards a volunteer cloud system," *Future Generation Computer Systems*, vol. 29, 2013, pp. 1442–1452.
29. B. Varghese and R. Buyya, "Next generation cloud computing: new trends and research directions," *Future Generation Computer Systems*, vol. 79, 2018, pp. 849–861.
30. S. Sahmim and H. Gharsellaoui, "Privacy and security in Internet-based computing: cloud computing, Internet of Things, Cloud of Things: a review," *Procedia Computer Science*, vol. 112, 2017, pp. 1516–1522.
31. A. Aliyu et al., "Cloud computing in VANETs: architecture, taxonomy, and challenges," *IETE Technical Review*, 2017, pp. 1–25.
32. A. Mayoral et al., "SDN orchestration architectures and their integration with cloud computing applications," *Optical Switching and Networking*, vol. 26, 2017, pp. 2–13.
33. J. D. Prince, "Introduction to cloud computing," *Journal of Electronic Resources in Medical Libraries*, vol. 8, no. 4, 2011, pp. 449–458.
34. A. Behl and K. Behl, "An analysis of cloud computing security issue," *Proceedings of World Congress on Information and Communication Technologies*, 2012, pp. 109–114.
35. D. R. R. Joan, "Encroachment of cloud education for the present educational institutions," *i-manager's Journal on Cloud Computing*, vol. 2, no. 2, February–April 2015, pp. 7–13.
36. S. Okai et al., "Cloud computing adoption model for universities to increase ICT proficiency," *Sage Open*, vol. 4, no. 3, July–September 2014, pp. 1–10.
37. L. S. Aaron and C. M. Roche, "Teaching, learning, and collaborating in the cloud: applications of cloud computing for educators in post-secondary institutions," *Journal of Educational Technology Systems*, vol. 20, no. 2, 2011–2012, pp. 95–111.
38. T. S. Behrend et al., "Cloud computing adoption and usage in community colleges," *Behaviour & Information Technology*, vol. 30, no. 2, 2011, pp. 231–240.
39. D. S. Linthicum, *Cloud Computing and SOA Convergence in Your Enterprise*. Upper Saddle River, NJ: Addison-Wesley, 2010.
40. J. Rhoton, *Cloud Computing Explained*. Dublin: Recursive Press, 2010.
41. J. Rosenberg and A. Mateos, *The Cloud at Your Service*. Greenwich, CT: Manning Publications, 2011.
42. B. Sosinsky, *Cloud Computing Bible*. Indianapolis, IN: Wiley Publishing, 2011.
43. S. Srinivasan (ed.), *Security, Trust, and Regulatory Aspects of Cloud Computing in Business Environments*. Hershey, PA: Information Science Reference, 2014.
44. S. Murugesan and I. Bojanova, *Encyclopedia of Cloud Computing*. Wiley-IEEE Press, 2016.
45. B. Furht and A. Escalante (eds.), *Handbook of Cloud Computing*. New York: Springer, 2010.

5

Cybersecurity

To steal from one writer is plagiarism, to steal from many is research, and to be inspired by any and all writers is creativity.

Anonymous

5.1 Introduction

We are all connected to the Internet one way or the other. The Internet has revolutionized the functioning of educational systems, businesses, economies, societies, and governments around the world. Although the Internet brings immeasurable opportunities, it also brings new risks. For all its advantages, increased connectivity brings increased risk of theft, fraud, and abuse.

The Internet is becoming a mess, open to all kinds of users, misuses, terrorists, spies, and identity thieves. Our major concern now is weapons of mass disruption, not weapons of mass destruction. We are at risk because America increasingly relies on computers and computer networks. Billions of machines—tablets, smartphones, ATM machines, smart homes, environmental control systems, thermostats, and much more—are all linked together.

By nature, cyberspace or the Internet is difficult to secure. Intruders exploit the vulnerabilities to steal information and money and perpetrate crimes. The crimes include child pornography, banking and financial fraud, and intellectual property violations. They may also include accessing government and defense confidential information, tampering with commercially sensitive data, and targeting supply chains. Companies are constantly bombarded from all types of sources: criminal syndicates, cyber vandals, intruders, and disgruntled insiders/employees.

The act of protecting information systems is known as cybersecurity. It is becoming more and more important as more information is being made available on computer networks. Cybersecurity is a national and global phenomenon because the malicious use of cyberspace could hamper economic, public health, safety, and national security activities. It takes

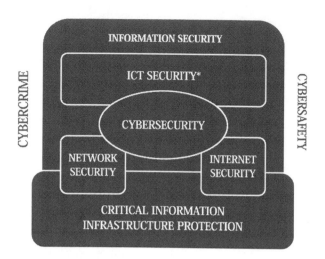

FIGURE 5.1
Relationship between cybersecurity and other security domains [1].

different forms including military, law enforcement, judicial, commerce, infrastructure, interior, intelligence, and information systems. The relationship between cybersecurity and other security domains is shown in Figure 5.1 [1]. Cybersecurity issues are critical for information infrastructure such as a smart power grid. The current trend of integrating power systems with advanced communication technologies has introduced serious cybersecurity concerns.

This chapter begins with cybersecurity features. It covers various forms of cyber threats: cyberattacks, cyberterrorism, and cyberbullying. It addresses cybersecurity governance and the trade-offs between cybersecurity and freedom. It further discusses cybersecurity applications, strategies, and challenges. Finally, the last section provides the conclusions.

5.2 Cybersecurity Features

The cybersecurity is a dynamic, interdisciplinary field involving information systems, computer science, and criminology. The security objectives have been availability, authentication, confidentiality, nonrepudiation, and integrity [2]:

- *Availability.* This refers to availability of information and ensuring that authorized parties can access the information when needed. Attacks targeting availability of service generally leads to denial of service.

- *Authenticity.* This ensures that the identity of an individual user or system is the identity claimed. This usually involves using username and password to validate the identity of the user. It may also take the form of what you have such as a driver's license, an RSA token, or a smart card.

- *Integrity.* Data integrity means information is authentic and complete. This assures that data, devices, and processes are free from tampering. Data should be free from injection, deletion, or corruption. When integrity is targeted, nonrepudiation is also affected.

- *Confidentiality.* Confidentiality ensures that measures are taken to prevent sensitive information from reaching the wrong people. Data secrecy is important especially for privacy-sensitive data such as user personal information and meter readings. Confidentiality is roughly the same as privacy. It is the assurance that information is shared only among authorized individuals. For IT, confidentiality is king.

- *Nonrepudiation.* This is an assurance of the responsibility to an action. The source should not be able to deny having sent a message, while the destination should not deny having received it. This security objective is essential for accountability and liability.

Information technologies have become indispensable to the modern lifestyle and threats against the availability, integrity, confidentiality, and nonrepudiation of information can affect the very functioning of our societies.

5.3 Cyberattacks

Cybersecurity is the process of protecting computer networks from cyberattacks or unintended unauthorized access. Cyberattacks are threatening the operation of businesses, banks, companies, and government networks. They vary from illegal crime of individual citizen (hacking) to actions of groups (terrorists). The following are typical examples of cyberattacks or threats [3]:

- *Malware.* This is a malicious software or code that includes traditional computer viruses, computer worms, and Trojan horse programs. Malware can infiltrate your network through the Internet, downloads, attachments, email, social media, and other platforms. Spyware is a type of malware that collects information without the victim's knowledge.

- *Scareware.* This is a malicious computer program that is meant to convince the victim that their system is infected, pressuring the victim to buy and download fake antivirus software. The protection software regularly displays warnings for infections and demands payment for removing them.

- *Phishing.* Criminals trick victims into handing over their personal information such as online passwords, social security number, and credit card numbers. Phishing seems to be the most common type of social engineering attack. It is associated with fake emails and websites. Phishing occurs when a malicious party sends a fraudulent email. The email is meant to trick the recipient into sharing personal information such as credit cards, passwords, or social security numbers. It may also involve enticing a victim to download an attachment or click a hyperlink. People sometimes will divulge sensitive or private information to those they feel obligated. Phishing has been around for a long time, but it has become more numerous and sophisticated.

- *Pretexting.* This occurs when one party lies to another in order to gain access to privileged information. An impostor creates a setting designed to influence the victim to release sensitive information. While phishing emails use fear and urgency to their advantage, pretexting attacks rely on building a false sense of trust with the victim. For example, the attacker may pretend the need some personal information in order to confirm the identity of the target.

- *Denial-of-service attacks.* These attacks generally involve thousands of computers, which generally overload a network resource with requests for information. They are designed to make a network resource unavailable to its intended legitimate users. These can prevent the user from accessing email, websites, online accounts, or other services.

- *Social engineering attacks.* Social engineering may be regarded as an umbrella term for computer exploitations that employ a variety of strategies to manipulate a user. It represents a real threat to individuals, companies, organizations, and governments. A cybercriminal attempts to trick users to disclose sensitive information. A social engineer aims to convince a user through impersonation to disclose secrets such as passwords, card numbers, or social security number. The variety of social engineering attacks is limited only by the human imagination [4,5].

- *Man-in-the-middle attack.* This is a cyberattack where a malicious attacker secretly inserts himself/herself into a conversation between two parties who believe they are directly communicating with each other. A common example of man-in-the-middle attacks is eavesdropping. The goal of such an attack is to steal personal information.

Cybersecurity threats or attacks are real and they happen to individuals in all walks of life on a regular basis. It has now become a national imperative and a government priority. Smaller companies and organizations are prime targets for hackers and malware, because these businesses lack cybersecurity plans and the resources.

Cybersecurity involves reducing the risk of cyberattacks. It involves the collection of tools, policies, guidelines, risk management approaches, and best practices that can be used to protect the cyber environment and mitigate cyberattacks. Cybercrime prevention is a multifaceted issue. Cyber risks should be managed proactively by the management.

5.4 Cyberterrorism

Cyberterrorism is the term used for terrorists who use Internet to communicate and wreak havoc and paralyze nations. It now ranks with other weapons of mass destruction and weapons of mass disruption in the public awareness. Although the U.S. has experienced hundreds of cyberattacks, none of them rose to the level of cyberterrorism. The U.S. Department of Homeland Security (DHS) recognizes that cyberterrorism is a serious issue. Although cyberterrorism is yet to occur in the U.S., terrorists groups in Sri Lanka and Japan have used cyberattacks against their own governments [6].

Cyberterrorism is the convergence of terrorism and cyberspace. It involves the use of the Internet as both enabler and support mechanism. It has the potential of creating a postmodern state of chaos. It uses computer resources to intimidate, harm or disrupt critical infrastructures such as power grid, transportation, oil and gas, banking and finance, water, and emergency services. Cyberterrorists could launch an attack on hospitals, destroying lives, their peacemakers, and life-support machines. They could also attack the military, destroying their communications systems. If the target is not critical infrastructure, the attack is not regarded as cyberterrorism. In order to qualify as cyberterrorism, it must occur in cyberspace and use some computer network to carry out the attack. For example, Al-Qaeda terrorist groups, who were responsible for the terror attacks of September 11, 2001, employed computer networks in their operations and caught the US by surprise.

Government officials and other professionals often claim that the world is unprepared for cyberterrorism. But what can be done to stop it? Preventives should be effective in blocking many forms of cyberterrorism. Critical networks, such sensitive military systems and computer networks for CIA and FBI, are not physically connected to the Internet, making them immune to cyberattack.

With the growth in cyber usage, cybercrime is challenging law enforcement agencies. Although crimes are unavoidable in modern world, they can be minimized through laws, regulations, law enforcement agencies, international cooperation, etc.

One challenge is defending all critical infrastructures. Traditional security strategies, such as encryption, firewalls, and contingency planning, are not enough to defend critical infrastructure. Computing professionals need to be aware of the whole gamut of vulnerabilities. An effective defense will require a concerted effort from the government and business executives. Fight against cyberterrorism is a continuous struggle.

5.5 Cyberbullying

Cyberbullying refers to a form of harassment that occurs online. Online bullying activities are common among adolescents and minors. It ranges from false accusations to obscenities. Its impact on the victim may include fear, depression, sadness, suicidal thought, embarrassment, trauma, helplessness, and worry. Cyberbullying can be prevented through the joint effort of students, parents, teachers, and the communities [7].

Cyberbullying is an aggressive, intentional behavior carried out by an individual or group using electronic technology. It is becoming common among teenagers. Tech-savvy students are taking advantage of the Internet and cell phones to bully their peers. The reason why they bully includes breakup in relationships (between boyfriends and girlfriends), rejection, retaliation/revenge, and desire to harm relationships.

The United Nations Convention on the Rights of the Child requires that adults protect children from violence, abuse, or injury. Federal authorities usually lead the way when it comes to enforcing laws against cybercrime. Many organizations provide resources on how to prevent and detect cyberbullying. Most of those resources are available at [8].

5.6 Cyber Espionage

Cyber espionage is stealing or spying using the Internet. The goal is typically to acquire intellectual property of government and private industry. Cyber espionage may be organized by foreign intelligence services (governmental espionage) or by corporations (industrial espionage) and should be regarded a significant part of cyber warfare. It has become the fifth domain of warfare, after land, sea, air, and space. There are five leading harmful activities in cyberspace are as follows: cybercrime, cyberattacks, cyber espionage, cyber terrorism, and cyber bullying. Any threat to our economy is a threat to our national security. One way to attack our economy is through cyber espionage activities and intellectual property theft. Such an attack has the potential to touch every person, whether military, civilian, business entity, or nation state.

One of the most serious threats to a modern economy is cyber espionage and insiders. As insider threat is increasing, so too is the threat from cyber spies. Due to the rapid growth in computerization and the drastic development of the Internet, cyber espionage has become the tool of choice for corporate and government spies. This new form of espionage is affecting the economic and political relationships between nation-states. Now most of the crucial and sensitive information, classified information, and top secrets are stored on computers and electronic devices.

The key feature of cyber espionage is that it occurs in secret, behind the scenes. Cyber espionage is specifically targeting secret information for malicious purposes. The cyber espionage activities are directed against high-tech industries and institutions such as telecoms, biotechnics, healthcare, and defense. One government aggressively spying on another, military tactfully send spies to other nations, and business spies on competitors are common. We live in a time when state-sponsored espionage continues to mount as a means of intelligence gathering.

Hackers or unknown intruders can install spyware on the computers to steal sensitive data. Cyber terrorists can attack you from any place at any time if your computer is not properly secure. You need to be always ahead of the game. It is important that you will have to take the following actions [9]:

1. *Have a data policy.* Make sure that only those who are allowed to access to critical information get access to it. Protect your critical infrastructure. Separate the network with the intellectual property from the corporate network.

2. *Data encryption.* Encrypting all sensitive data at rest and in transit can make it more difficult for hackers to intercept and decipher.

3. *Bring your own device* (BYOD). This may cause you more problems than it is worth if not properly controlled. Have some kind of device control mechanisms in place that will safeguard against data leakage.

4. *Monitor for unexpected activity or behavior.* The reaction to cyber espionage activities varies around the world. Microsoft sent a security breach report and released a security patch. An attacker may try to sidestep technical security mechanisms by offering a bribe to a project team member. Employees may readily fall into the temptation of stealing information for monetary rewards [10].

5. *Deterrence.* Deterrence is the proper tool for nation-state cyber espionage. It occurs when a nation convinces its enemy that it is willing and able to respond to cyber intrusions using military force. This is a useful counter-espionage strategy when a nation has the resources to carry it out.

Cyber espionage is one of the most intriguing international problems today. It is changing the shape of modern warfare. There are two major trends associated with modern nation-state cyber espionage. The first is that cyber espionage is becoming more advanced, effective, and professional. The second trend is that cyber espionage is becoming an accepted, and even preferred means of warfare [11].

5.7 Industrial Espionage

It is widely recognized that information is a golden resource with high economic value. Every business thrives on information. Information can make the difference between success and failure of the business. Cheating and stealing information are natural consequences of competition and information economy, which relies on technology and the Internet. Industrial or corporate espionage is obtaining information from a rival company to sabotage its operation. Industrial espionage is often called economic espionage in order to distinguish it from more traditional forms of national security espionage. It usually occurs between companies or corporations. It may also involve covert activities such as theft, bribery, blackmail, and technological surveillance. American business executives are concerned or worried about industrial espionage because it is a major source of loss. In the high-tech world, no business is free from industrial espionage.

The main purpose of industrial espionage is to spy and gather information about an organization or company. The information may include trade secrets, patents, designs, ideas, techniques, processes, recipes, formulas, pricing, sales, marketing, or any other sensitive data. Industrial espionage is commonly associated with technology-heavy industries including automobiles, aerospace, telecommunications, biotechnology, and energy [12].

An example of industrial espionage in history involved the attempt by Europeans to acquire the secret of China's porcelain manufacturing process. Another example involved the Soviet industrial espionage. The Soviets built the Tupelov supersonic aircraft which closely resembled the Concorde aeroplane, built by Britain and France. Through spying activities the Soviets sought to develop their own microelectronics, but their technology appeared to be several years behind the West's. The political cold war may be over but the economic cold war seems to be ongoing. The Chinese, Russians, Israelis, Japanese, Germans, and South Koreans have also been accused of industrial espionage.

Capitalism causes intense competition between companies competing for a bigger share of the market. Since information can make the difference between success and failure, some businesses will go to any length to get ahead, even if it means spying. Common types of industrial espionage include the following: (1) *Social engineering.* This kind of attacks often involves employees of one company attempting to infiltrate a rival company. (2) *Hacking.* This involves breaking into a computer system and stealing the information on the computer via the Internet. (3) *Dumpster diving.* This involves looking through the garbage of a rival company and looking for any important information that may have been thrown away.

Although espionage activities are most often unethical and illegal, such covert activities have led to the advancement of technology in many sectors and many nations. They save companies time as well as huge sums of money they could have spent on R&D. Thus, industrial espionage has been an illegal form of technology transfer. It has been argued that the only way that China could have developed as rapidly as it has is through its availing itself of massive amounts of foreign technology.

Sometimes, industrial espionage can be legal. Competitors may buy products from a company and have a working technology to explore. Reverse engineering involves dismantling the target piece of technology and then reproducing it. A company may buy another company to acquire its core technology [13]. In order for a company to do business in the foreign land, the company must train native workers in a critical technology. There is a cost/benefit analysis that needs to be done before entering into such a venture. The hiring away of employees may also lead to the transfer of knowledge to a competitor [14].

Industrial espionage is difficult to prove because it is often performed by insiders that already have access to sensitive data. It is almost indistinguishable from their normal everyday activities. Such actions are very hard to detect and even harder to prove in court. All organizations gather and make use of some kind of information about their competitors. But it is difficult to determine at what point legitimate competitive intelligence gathering crosses the line into industrial espionage [15]. The US enacted the Economic Espionage Act (EEA) in 1996 to deal with espionage and criminalize stealing intellectual property. The act provides a powerful tool for law enforcement to protect companies while punishing spies [16]. However, since laws on industrial espionage are different from nation to nation, it may be very hard to hold foreign perpetrators accountable.

5.8 Cybersecurity Governance

Cybersecurity is the joint responsibility of all relevant stakeholders including government, business, infrastructure owners, and users. Governments and international organizations play a key role in cybersecurity issues. Securing the cyberspace is of high priority to the DHS. The DHS has a dedicated division responsible for risk management program and requirements for cybersecurity called the National Cyber Security Division. The Federal

Communications Commission's (FCC) role in cybersecurity is to strengthen the protection of critical computer networks and networked infrastructure. The Computer Fraud and Abuse Act (CFAA) remains the most relevant applicable law expressing the U.S. proactive cybersecurity effort.

Other governments (such as the United Kingdom, Canada, Japan, Australia, and New Zealand) are introducing various security measures, enacting cybersecurity related laws and regulations, and forging international cooperation on cybersecurity. Cybersecurity issues have been on the NATO agenda for a while due to its international nature. NATO redefined its cyber defense policy with its 2008 response to cyberattacks against Estonia [17].

5.9 Cybersecurity and Freedom

Experts and policymakers have weighed trade-offs between cybersecurity and freedom, often seeking to ensure both. Some countries boast free Internet, while others impose strict censorship and may punish citizens for what they post. Increasing cybersecurity will help protect consumers and businesses and ensure the availability of critical infrastructures on which our economy depends.

The United States' domestic telecommunications regulator, the FCC, has promoted network neutrality and uncensored Internet. Our right to private communications is a cornerstone of American democracy. But due to heightened awareness of terrorist attacks, new laws and policies introduced in the last decade have eroded our civil liberties online. This allows Internet Service Providers (such as AT&T, Comcast, and Verizon) to arbitrarily block potentially dangerous websites. When the government intends to protect its own systems, privacy issues may be raised [18,19].

Efficient cybersecurity can help protect privacy, but information that is shared to assist in cybersecurity efforts might sometimes contain personal information which some would regard as private.

5.10 Cybersecurity Applications

Here we consider some useful applications where cybersecurity is required [20]:

1. *Online identity theft.* This involves copying another person's identifying information (such as their name and Social Security Number) and then impersonating that person to perpetrate fraud or other

criminal activity. Online attackers attempt to steal users' identity during a bank or commercial transaction.

2. *Industrial attacks.* Attackers seek valuable intellectual properties of companies stored in corporate networks. Cyberattacks on the industry represent a threat beyond the boundaries of the factory involved. The consequences of cyber security incidents are diverse and costly. For chemical manufacturers, for example, the consequences can range from production interruption, reputation loss, supply chain impact, and the expense of retrofitting security after an incident.

3. *Critical infrastructure.* Cybersecurity risk pervades all sectors of the US economy. Terrorists can wipe out our power grid, telecommunication infrastructure, or banking system. Terrorism challenges the reliability, resiliency, and safety of our infrastructures. The chemical industry, particularly the petro-chemical industry, is a critical infrastructure that is vulnerable to cyberattacks. Our adversaries are after destroying our critical infrastructures such electric power grid, chemical refineries, and communications systems. Security issues arise as smart power grids (which allow a bidirectional flow of information) become targets of cybersecurity threats. With the rapid expansion of the Internet of Things (IoT), the potential for malicious attacks on the smart grid is on the increase. Advanced network security in the form of intrusion detection system (IDS) and intrusion prevention system (IPS) will protect smart grids from many of the more advanced and emerging cyber threats [2]. There is a potential cyber threat to supervisory control and data acquisition (SCADA) systems, a type of industrial control systems, which control and monitor our critical infrastructures such as energy, waste and water systems, telecommunications, transportation, emergency services, financial services, commercial facilities, defense, food and agriculture, healthcare and public health, and nuclear reactor. These critical infrastructures are interdependent and their disruption could be catastrophic; it could affect global critical infrastructure through a cascade effect. Cyberattacks on health and business sectors and government agencies have grown exponentially.

5.11 Cybersecurity Strategies

In spite of government's involvement in cybersecurity, it is the duty of a company to implement the measures and strategies needed to mitigate cyber risks. Each company should have established a cybersecurity policy and implement a real strategy regarding cybersecurity. The following guides are offered for mitigating cybersecurity risks.

- *Detecting instructions.* The network systems will have to be able to optimize their network detection. It should prevent against data leakage using defensive strategies that prevent breaches of security. A system should always be protected from unwelcome visitors accessing it. It is highly recommended that a single sign on and password is given for each user to access all they need. A defense strategy must take a multi-layered approach to cybersecurity. No single security measure is good enough to prevent intrusions. A defense-in-depth strategy involves layers of protection on assets, intrusion detection, and continuous monitoring, as shown in Figure 5.2 [21].

- *Supporting cybersecurity improvements.* The proactive support of cybersecurity in the long run will require a strong and lasting commitment of resources, some clear goals, and close collaboration between the sector's stakeholders. Chemical industry trade associations have launched different cybersecurity programs to help their members improve chemical industry cybersecurity, comply with federal regulations, and mitigate vulnerabilities to cyberattacks. This also applies to other industries.

- *Protecting operations.* To make the industry more resilient to cyberattacks and better able to protect our interests in cyberspace, the

FIGURE 5.2
Defense-in-depth approach to security [21].

industry must create a culture for security through ongoing awareness campaigns that will protect the business. Security requires an ongoing effort and must be continually managed through the lifecycles of any company. Engineers must ensure that availability, integrity, and confidentiality (AIC) are fully implemented within the overall system approach.

- *Regulation.* Regulations should continue to require facilities to meet stringent security standards. To secure industrial facilities, the federal government must provide regulatory certainty. DHS should address this issue and ensure that all high-risk chemical facilities are safe, secure, and fully comply with CFATS. A variety of federal policy options can be used for security.

- *Physical security enhancement.* Site security can be improved by "hardening" defenses so that sites would be less vulnerable to terrorists. Protecting the chemical facilities, information regarding chemical formulas, and customer databases, from potential cyber-attacks, is crucial.

- *Firewall.* One of the key methods of prevention is to maintain up-to-date firmware on all security devices. A firewall may serve as an essential defense against unsolicited Internet traffic coming or going from your computer. A firewall is a type of security barrier placed between network environments. Only authorized traffic is allowed to pass.

A proactive approach to dealing with cybersecurity attacks must involve a cost-effective and realistic approach that works for the employers and employees.

5.12 Challenges

Cybersecurity policy faces a host of challenges and obstacles—political, bureaucratic, legal, political, financial, national, and international. Unfortunately, we lack proper scientific understanding of cybersecurity in order to tackle these challenges in a principled manner.

The first challenge cybersecurity faces is the confusion over its varied definitions [22]. Cybersecurity means different things to different people and there is lack of consensus among stakeholders. There is no uniform set of standards that cybersecurity professional must follow when implementing a cybersecurity program. Although there are several laws, regulations, and policies governing cybersecurity, compliance with such guidelines does not automatically guarantee network security.

Cybersecurity policy lags technological innovation. Information technology changes rapidly, with security technology and practices evolving even faster to keep pace with changing threats. Because cybersecurity is not well-understood by non-experts, the economics are hard to demonstrate, and effectiveness is difficult to measure. Minimizing our cybersecurity risks requires commitment on both technical and political fronts.

The security challenge facing most industries today is twofold: designing new systems that meet cybersecurity standards and modifying existing systems to meet cybersecurity requirements.

Security is no longer confined just to what users do with computer. Through the IoT, devices can gather, store, and transmit data that are capable of exposing sensitive information. The explosion in these and inherently insecure devices is shifting the security paradigm. IoT manufacturers must improve the security of their devices because good cybersecurity is good business [23].

Mobile technology and social networking brings new cybersecurity challenges. With the advent of such applications at an unprecedented scale, the privacy of the information is compromised to a larger extent.

Cybersecurity breaches cannot be stopped at a nation's borders since it is difficult to determine where the actual borders are in cyberspace. Thus, cybersecurity can become a supranational problem. To address cybersecurity threat, nations must collaborate among themselves [24].

5.13 Conclusion

Cybersecurity has now become a national imperative and a government priority. It is one of the main concerns of many organizations today. The US government, at the state and federal levels, has recognized the growing importance of securing the US cyberspace and is committed to prosecuting cybercrimes and holding those accountable for perpetrating acts. Private sector, however, is still responsible for the security of their private networks. Preserving cybersecurity is a difficult problem and there is no obvious solution to the problem of cybersecurity. Reducing our cyber risk requires progress on both technical and political fronts [25].

Cybersecurity education and training are crucial toward protecting the nation's ever increasing cyberinfrastructure. One way to do this is to integrate cybersecurity concepts to undergraduate programs in STEM, computer science, and IT fields, which already offer the core technology skills [26]. Some contend that cybersecurity should be a discipline by itself, and not merely adding one or two courses to the curriculum. A growing number of colleges globally offer such programs [27].

The demand for cybersecurity professionals is expected to grow since weak cybersecurity endangers the country. Companies, government, and organizations all employ cybersecurity professionals [28].

References

1. A. Klimburg (ed.), *National Cyber Security Framework Manual*. Tallinn: NATO CCDCOE Publication, 2012.
2. M. N. O. Sadiku, M. Tembely, and S. M. Musa, "Smart grid cybersecurity", *Journal of Multidisciplinary Engineering Science and Technology*, vol. 3, no. 9, September 2016, pp. 5574–5576.
3. FCC small Biz cyber planning guide, https://transition.fcc.gov/cyber/cyberplanner.pdf.
4. M. N. O. Sadiku, A. E. Shadare, and S. M. Musa, "Social engineering: an introduction", *Journal of Scientific and Engineering Research*, vol. 3, no. 3, 2016, pp. 64–66.
5. J. M. Hatfield, "Social engineering in cybersecurity: the evolution of a concept", *Computers & Security*, vol. 73, 2018, pp. 102–113.
6. M. N. O. Sadiku, A. F. Shadare, S. M. Musa, and C. M. Akujuobi, "Cyberterrorism", *Journal of Multidisciplinary Engineering Science and Technology*, vol. 3, no. 12, December 2016, pp. 6269–6271.
7. M. N. O. Sadiku, A. E. Shadare, and S.M. Musa, "Cyberbullying: a primer", *International Journal of Advanced Research in Computer Science and Software Engineering*, vol. 7, no. 3, March 2017, pp. 298–300.
8. A. R. Hathcote and K. A. Hogan, "Resource guide on cyberbullying", *Preventing School Failure*, vol. 55, no. 2, 2011, pp. 102–104.
9. S. Sharma, "Cyber espionage: The spy who hacked me", http://www.pcquest.com/cyber-espionage-spy-hacked/.
10. A. Laszka, et al., "Secure team composition to thwart insider threats and cyber-espionage", *ACM Transactions on Internet Technology*, vol. 14, no. 2–3, October 2014.
11. D. Rubenstein, "Nation state cyber espionage and its impacts," http://www.cse.wustl.edu/~jain/cse571-14/ftp/cyber_espionage/.
12. "Industrial espionage," *Wikipedia*, the free encyclopedia, https://en.wikipedia.org/wiki/Industrial_espionage.
13. H. J. Bhatti and A. Alymenko, "A literature review: Industrial espionage," https://www.researchgate.net/publication/318816635_A_Literature_Review_Industrial_Espionage.
14. I. S. Winkler, "Case study of industrial espionage through social engineering," https://csrc.nist.gov/csrc/media/publications/conference-paper/1996/10/22/proceedings-of-the-19th-nissc-1996/documents/paper040/winkler.pdf.
15. A. Crane, "In the company of spies: When competitive intelligence gathering becomes industrial Espionage," *Business Horizons*, vol. 48, 2005, pp. 233–240.

16. T. R. Stutler, "Stealing secrets solved: Examining the Economic Espionage Act of 1996," *FBI Law Enforcement Bulletin*, vol. 69, no. 1, November 2000, pp. 11–16.
17. E. Tikk, "Global cybersecurity–Thinking about the niche for NATO," *SAIS Review*, vol. 30, no. 2, Summer/Fall 2010, pp. 105–119.
18. G. T. Nojeim, "Cybersecurity and freedom on the Internet," *Journal of National Security Law & Policy*, vol. 4, 2010, pp. 119–137.
19. M. Ammori and K. Poellet, "Security versus freedom' on the internet: cybersecurity and net neutrality," *SIAS Review*, vol. 30, no. 2, Summer/Fall 2010, pp. 51–65.
20. T. Moore, "The economics of cybersecurity: principles and policy options," *International Journal of Critical Infrastructure Protection*, vol. 3, 2010, pp. 103–117.
21. A. Ginter and W. Sikora, "Cybersecurity for chemical engineers," *Chemical Engineering*, vol. 118, no. 6, June 2011, pp. 49–53.
22. K. Zeng, "Exploring cybersecurity requirements in the defense acquisition process," *Doctoral Dissertation*, Capitol Technology University, April 2016.
23. K. L. Miller, "What we talk about when we talk about 'reasonable cybersecurity': A proactive and adaptive approach," *The Computer & Internet Lawyer*, vol. 34, no. 3, March 2017, pp. 1–8.
24. H. de Bruijn and M. Janssen, "Building cybersecurity awareness: the need for evidence-based framing strategies," *Government Information Quarterly*, vol. 34, 2017, pp. 1–7.
25. S. L. Garfinkel, "Inside risks: the cybersecurity risk," *Communications of the ACM*, vol. 55, no. 6, June 2012, pp. 29–32.
26. D. Manson and R. Pike, "The case for depth in cybersecurity education," *ACM Inroads*, vol. 5, no. 1, March 2014, pp. 47–52.
27. K. J. Knapp, C. Maurer, and M. Plachkinova, "Maintaining a cybersecurity curriculum: professional certifications as valuable guidance," *Journal of Information Systems Education*, vol. 28, no. 2, December 2017, pp. 101–113.
28. "Computer security," *Wikipedia*, the free encyclopedia, https://en.wikipedia.org/wiki/Computer_security.

6

Software-Defined Networking

Success is not final, failure is not fatal: it is the courage to continue that counts.

Winston S. Churchill

6.1 Introduction

Although there have not been major changes in traditional networks since 1970, the complexity and rigidity in current traditional IP networks have made them difficult to manage. Managing such a huge complex network is a big challenge and is prone to errors. This is largely due to fact that control and data planes are vertically integrated, rigid, and vendor specific. For example, setting up a network requires multiple software and hardware. Devices (such as switches, routers, gateways) must be configured individually making it time consuming to update. These devices have a static architecture. As the network increases in size, its complexity grows drastically [1,2].

The needs of business have surpassed the networks' ability to provide service. The rising demands for mobile devices, cloud-based mobility, social media, server virtualization, social networking, multimedia, and big data (more bandwidth) services are pushing traditional networks to their limits and making them to be ill-suited to meet the requirements to today's enterprises, carriers, and end users. A new approach is required. Software-defined networking (SDN) provides that approach.

SDN addresses the failure of the traditional networks to support the dynamic, scalable computing, and storage needs of today's applications. SDN achieves this by separating or decoupling network control from data forwarding, with the aim of increasing network agility, network programmability, and easy management [3]. Figure 6.1 compares traditional and SDN architectures [4]. SDN is complemented by virtualization technologies such as network function virtualization (NFV).

This chapter begins with the historical background on SDN. It covers SDN architecture and features. Then it covers various applications and benefits of SDN. It presents the challenges hindering wide adoption of SDN. It finally presents security as the major concern of researchers from academia and industry.

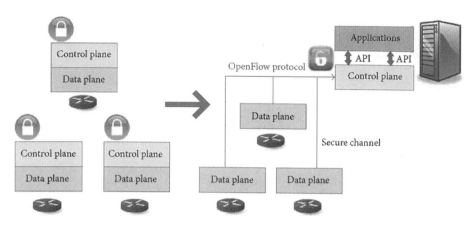

FIGURE 6.1
A comparison of traditional and SDN architectures [4].

6.2 Historical Background

The idea of SDN came from Stanford University's project, named OpenFlow in 2006 by Nick Mckeown, a professor of electrical and computer science. The term "software-defined networking" was first used by Open Networking Foundation (ONF), which is a nonprofit industry consortium founded in March 2011. It was formed to accelerate the development, standardization, and commercialization of SDN. The consortium consists of 69 members including IBM, Cisco, Juniper, HP, and Dell. The goal of ONF is to transform networking industry to software industry through SDN and promote standard development. ONF promotes the use of SDN and OpenFlow protocol. The OpenFlow is a standard interface between the control and data planes. OpenFlow 1.0 was released in December 2009. Major commercial switch vendors such as IBM, HP, and Cisco have launched switching products that support the OpenFlow protocol. Google has deployed SDN in its data centers across the globe.

6.3 SDN Architecture

SDN is a new computer networking architecture that uses standardized application programming interface. In traditional networks, most of the network functionality is in form of hardware such as like switches, routers, and gateways. When a new technology is to be introduced, all hardware devices must be replaced. Traditional networking architectures are not dynamic and have limited functionality, which must be overcome to meet today's requirements. SDN promises to remove the limitations on current network infrastructure.

SDN is the next wave of networking. It simplifies the traditional networking in two ways. First, the network now consists of uniform switching hardware with standard interfaces. Second, network control is not distributed but centralized and restricted to the controller. The basic SDN architecture [5] is shown in Figure 6.2. It consists of three planes: data, control, and application, with standardized interfaces between control and data planes [6]:

- *Data (or infrastructure) plane.* The data plane is the bottom plane and has dedicated hardware devices such as routers, switches, and access points. It is responsible for processing packets and forwarding information to the upper controllers. Thus, packet forwarding (using switches) is one of the basic functions of the data plane. A software-centric data plane consists of software-based packet switching and forwarding.
- *Control plane.* The control plane manages all decision-making protocols and access-control algorithms. It is responsible for monitoring the network, making routing decisions, and programming

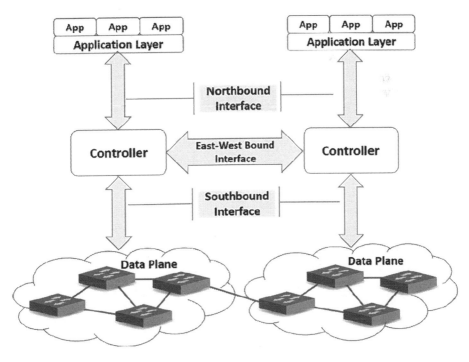

FIGURE 6.2
Basic SDN architecture [5].

the network. It consists of a set of software-based controllers (or network operating systems (NOS)), which may be regarded as the brain of SDN. The controllers interact with applications and devices through three application programming interfaces (to be mentioned later). An SDN controller uses a global view to make globally valid decisions. Programmability to configure the network is provided on the control plane. Network devices are used only for forwarding; decisions about where data is sent are taken by the controller.

- *Application plane.* The application layer lies at the top of the SDN architecture. All applications reside in this layer. The layer consists of the end-user applications (security, visualization, etc.) that utilize the SDN communications and network services that are designed to meet user demands. SDN applications are programs that explicitly and programmatically communicate their network requirements. Such applications provide routing, monitoring, security, load balancing, flow control, resource management, network virtualization, and traffic management.

In addition to the architecture of SDN, multiple application program interfaces (APIs) exist—*northbound, southbound, eastbound,* and *westbound.* SDN uses *southbound* APIs to relay information to the switches and routers "below." OpenFlow was the original southbound API. OpenFlow is an API that interfaces data and control planes. It is a protocol that describes how a controller communicates with network devices such as switches and routers. SDN uses *northbound* APIs to communicate with the applications and business logic "above." The *eastbound* and *westbound* APIs are responsible for interfacing between multiple controllers. How controllers interact with one another to share information is handled via their east–west interface.

6.4 SDN Main Features

SDN is characterized by six fundamental features: (1) separation of the data and control planes, (2) centralization, (3) programmability, (4) network virtualization, (5) resource abstraction, and (6) openness. These may be regarded as essential building blocks or pillars of SDN.

- *Plane separation.* The main advantage of SDN over traditional IP network is the separation of control plane and data plane, which results in centralized control [7]. This separation of the control and data planes allows one to experiment with network protocols. The decoupled networking architecture enables timely delivery of control message and impacts the efficiency of SDN. The data plane is responsible for the packet forwarding and the control plane performs other functions. In other words, the control plane makes decisions about where traffic is sent, while the data plane forwards traffic to the desired destination. The network switches/ routers forward packets by following the flow table rules determined by the control plane. The control plane is programmable and is implemented in a centralized mode. Various applications may run on top of the centralized controller. It is felt that the shift to a pure software model brings flexibility, efficiency, and agility unknown in traditional networks. Although SDN decouples the data and control planes, the performance of one plane affects the performance of the other. Thus isolation of the tenants' control plane is necessary. The performance of the control plane is affected by the resources available for the hypervisor (also known as virtual machine monitors) instances. The hypervisor performs two main functions: abstraction (virtualization) of the underlying physical SDN network and isolation of the virtual SDN [8].

- *Centralization.* Instead of a distributed control architecture, SDN consolidates all the control in a single node—the network controller, which is a software running on a commercial server platform. The software-based controller manages SDN network using higher-level policies. The controller communicates with the applications through the northbound interface and with switches through the southbound interface. Network intelligence and state are logically centralized. Centralization of control makes sensing and adjusting control much faster than with distributed control. A major advantage of centralized control is the state or policy changes propagate much faster than distributed system. However, having a centralized controller (for overseeing the operation of the entire network) implies that security attacks can be focused on that one point of failure.

- *Programmability.* This is one of the fundamental tenets of SDN. The network devices are programmable through software applications running on top of the NOS. SDN makes the network more programmable in that it allows various network functionalities to be implemented in software. This way it makes the network fully adaptable to the changing needs of the users, network operators, and the applications. Programmability and automation of network resources enables service providers unlock new revenue opportunities, adapt to real-time changes and reduce network complexity [9,10].

- *Network virtualization.* This is basic to SDN. Virtualization allows software to run separately from the underlying hardware. Network functions that are being performed by hardware devices (such as load balancers, firewalls, routers, switches, and gateways) are being virtualized as virtual appliances. Network virtualization allows isolated virtual networks (such as M2M, smart grid, etc.) to share the same physical network infrastructure. For cloud computing and data centers, network virtualization for multi-tenants is important because it offers better utilization of resources and faster turnaround times for creating a segregated network. Resource virtualization is important for five reasons [11]: sharing, isolation, aggregation, dynamics, and ease of management.

- *Abstraction.* An abstraction is a representation of an entity in terms of selected characteristics, while hiding others irrelevant to the selection criteria. Abstraction establishes common security models that can be deployed repeatedly. Resource abstraction allows control to be separated from network resources so that network providers can directly manage their own sources. Resources typically include storage, processing, forwarding, or anything that is needed to deliver a service. Abstraction enables building scalable, flexible, and modular network control systems. An SDN can have three basic abstractions [12]: forwarding, distribution, and specification. Forwarding abstraction allows any forwarding behavior demanded by the network application. Distribution abstraction makes distribution control centralized and shields SDN applications from the vagaries of distributed state. Specification abstraction allows a network application to express desired network behavior without being asked to implement the behavior itself.

- *Openness.* Openness and interoperability are essential to SDN solutions. SDN is an open technology, leading to greater interoperability and more innovation. Open SDN implies that its interfaces should remain standard, not proprietary. It should support all major server operating systems (OS) and hypervisor platforms such as Microsoft, VMware, and Linux [13]. Open interfaces are defined between devices in the data plane and those in the control plane. Open SDN and open interfaces allow equipment from different vendors to interoperate. Organizations develop open standards. These include OpenDaylight which is an open development initiative hosted by the Linux Foundation in 2013, Open Platform of NFV which is dedicated to accelerating the adoption of NFV elements, and OpenStack which is a cloud operating controller [14]. NFV was developed by service providers such as AT&T, BT, Verizon, Deutsche Telekom, Orange, and Telefonica. All these standardizations efforts are meant to be vendor independent.

6.5 SDN Applications

SDN has been shown to be valuable in many applications. It has attracted a lot of attention in fields such as 5G mobile networks, cloud computing, wireless networks, data centers, big data analytics, Internet of Things (IoT), and optical networks.

- *Wireless networks.* Most SDN solutions are based on wired networks. Recently, attempts have been made to adapt SDN to wireless networks. SDN provides a global centralized control of access points. An SDN controller can usually manage thousands of access points simultaneously. Several SDN controllers can be deployed when there are a large number of access points. Network administration in Wi-Fi is mostly centralized [15]. SDN concepts have been used to improve vehicular ad hoc networks (VANETs) and cellular networks. For example, the hidden terminal problem in wireless networks ceases to be an issue if transmission is centrally controlled [16].

- *Wireless sensor networks.* Wireless sensor networks (WSNs) have benefited from the SDN approach. A typical WSN consists of a base station, which has the SDN controller, and sensor nodes. The control plane is decoupled from the data plane which runs the sensor nodes. A centralized controller uses OpenFlow to interact with the nodes. The nodes are often low powered and small. They are autonomous and adaptive to their environment.

- *IoT.* Everything in the world is being connected by the IoT. IoT allows connected devices to be controlled and accessed remotely. Traditional network technologies cannot handle IoT in an efficient, scalable, seamless, and cost-effective manner. Developers use the abstractions in the application layer to build IoT applications. The integration of SDN schemes in IoT will result in an evolving scalable, energy-efficient, and cost-effective IoT architecture. SDN and network virtualization are the two key technologies that will enable IoT networks [17,18].

- *Optical networks.* Optical networks either maintain signals in the optical domain or use transmission channels that carry signals in the optical domain. Software-defined optical transceivers can be flexibly configured by SDN [19]. SDN can be extended to the optical transport networks that interconnect data centers. Transport SDN enables efficient load balancing among widely dispersed data centers through a centralized controller.

- *Data centers.* The advances in data centers have strained the capabilities of traditional networks to the breaking point. Today's mega data centers hold thousands of physical servers. The data centers can be classified into three categories [20]: private single-tenant, private multitenant, and public multitenant. The dynamic allocation of resources between tenants of the data center and the general public makes SDN attractive. Deploying SDN in a data center infrastructure will enable automated application-aware mapping of traffic flows. It will dynamically instantiate networks and disable them when they are not needed. It may lead to energy efficiency and robustness. SDN has been adopted by Internet companies like Facebook, Google, and VMware for their data center networks.

- *WAN and LAN.* Wide area networks (WANs) have been used to connect geographically dispersed local area networks (LANs). Software-defined mobile networking is proposed as an extension of SDN to incorporate mobile network functionalities. An SD-WAN is a software-defined WAN that uses the principles of SDN. A SD-LAN is a software-defined LAN is built on the principles of SDN [21].

Other applications of SDN include home networks, enterprise networks, 5G cellular systems, cyber-physical networks, mobile networks, intrusion detection, monitoring systems, next-generation networks such as the future Internet, real-time video applications such as video conferencing and distance learning, underwater communication systems.

6.6 SDN Benefits

The revolutionary concept of SDN has been proposed to solve ossifications of Internet. The main benefits of SDN are its programmability, openness, and agility. SDN offers other benefits including on-demand provisioning, automated load balancing, eliminating middleboxes (devices that manipulate traffic for reasons other than forwarding, such as a firewall and load balancer), and a centralized, flexible, scalable, resilient, programmable network that allows cloud computing and network administrators to respond quickly to changing business requirements. Flexibility is the system's capability to adapt to dynamic network settings, while scalability is the network's capability to handle a growing amount of work. The administrator can change any rules for network switch if necessary. SDN networks have more control capabilities than legacy networks and they deliver speed, agility, flexibility, and programmability. They provide central management and reduce the overall deployment cost.

   SDN has better traffic-handling capacity. In SDN, network administrators can actually shape network traffic from a centralized network console. They can also change data traffic rules on-the-fly if necessary. SDN is totally software-centric. A central OS controls every connected device. SDN removes manual configuration and hardware dependence. SDN enables the design of a future Internet.

6.7 SDN Challenges

Despite the proven benefits of SDN, there is significant reluctance in adopting it. Like any new technology, SDN has some disadvantages. These include controller reliability, resilience, redundancy, performance, scalability, and interoperability between devices from multiple vendors. Failure of SDN controller can result in single point failure. Split control architectures like SDN are often questioned regarding their being resilient to faults. Having multiple controllers improves reliability because the data plane continues to operate when one controller fails. Transition from traditional network to SDN will require interoperability of existing network equipment with SDN. Service providers operate networks that are frequently stressed with high traffic volumes, rich media content, and security considerations that SDN has not yet addressed. Many organizations do not have the resources (time, capital, expertise, etc.) to invest in a new networking architecture [22].

Since it is difficult to implement SDN on existing physical infrastructure (which will not disappear for years to come), new SDN infrastructure must be built from the ground up. In addition, there are issues with fault-tolerance, universal standards, and integration with existing network infrastructure [5]. Solution to these problems will require expertise from network and traffic engineers, reliability engineers, computer scientists, mathematicians, and network theoreticians.

6.8 SDN Security

Security is a major concern of researchers and industries for adopting SDN. It was not considered as part of the initial design of SDN. Security deals with preventing malicious attacks, coping with natural disasters, and guaranteeing that the data are only accessible by authorized users. Being dynamic in nature, SDN is prone to security attacks. SDN can be used to secure data offloading from mobile and handheld devices. While SDN can be leveraged to secure greater security for networks, it faces some major challenges securing itself. Due to the distinctive features of SDN, traditional network security approaches cannot be applied directly to it. The ability to programmatically control network behavior opens up possibilities for network security. Although it improves network performance, SDN creates some peculiar problems. The centralized control and programmability features of SDN introduce some new security challenges [23].

Security threats to SDN include spoofing, tampering, repudiation, and denial of service (DoS). For example, there is an increased potential for DoS attacks due to the centralized controller. The controller is the easiest target for DoS, which aims at hindering the availability of resources in SDN. OpenFlow is vulnerable to man-in-the-middle attacks when Transport Layer Security is not used. Network breaches may result when network controllers are shared by different users or applications [24,25].

The issues of confidentiality, integrity, privacy, and availability must be addressed to achieve secure SDN. It is important that SDN should be designed with security in mind right from the start. That implies that security issues should be identified and resolved to enable reliable wide area SDN deployment [26].

6.9 Conclusion

As new technologies, such as IoT, cloud computing, and content delivery networks emerge, the traditional network architecture becomes a handicap. The legacy network architecture cannot meet all the requirements of new applications. SDN is the next wave of networking, which decouples the control plane and the data plane thereby overcoming the challenges of traditional networks. It is an evolutionary paradigm that makes the network programmable, vendor neutral, and easily configurable. It represents the next generation of infrastructure automation that is completely programmable and application-aware. The networking industry is enthusiastically embracing SDN as the promising solution for future Internet.

SDN is an emerging, disruptive technology that is yet to be fully explored; a lot is still going on with standards development. It is an integral part of "software-defined everything" [27]. It is becoming increasing important for the next-generation networks such as Internet backbone and data center networks. It has been adopted by Internet companies like Facebook and Google for their data center networks.

To raise modern networking literacy among students, courses are being offered on SDN [28]. More information on SDN can found in comprehensive surveys on the subject [2,5,10,12–15,29–33].

References

1. M. F. Tuysuz, Z. K. Ankarali, and D. Gozupek, "A survey on energy efficiency in software defined networks," *Computer Networks*, vol. 113, 2017, pp. 188–204.
2. R. Masoudi and A. Ghaffari, "Software define networks: a survey," *Journal of Network and Computer Applications*, vol. 67, 2016, pp. 1–25.

3. M. N. O. Sadiku, M. Tembely, and S. M. Musa, "Software-defined networking concepts," *Journal of Scientific and Engineering Research*, vol. 3, no. 5, 2016, pp. 92–94.

4. A. L. Valdivieso et al., "SDN: evolution and opportunities in the development IoT applications," *International Journal of Distributed Sensor Networks*, vol. 10, no. 5, 2014, pp. 1–10.

5. J. H. Cox et al., "Advancing software-defined networks: a survey," *IEEE Access*, vol. 5, 2017, pp. 25487–25526.

6. D. B. Hoang and M. Pham, "On software-defined networking and the design of SDN controllers," *Proceedings of the 6th International Conference on the Network of the Future*, Montreal, Canada, September/October 2015.

7. J. Xie et al., "Control plane of software defined networks: a survey," *Computer Communications*, vol. 67, 2015, pp. 1–10.

8. A. Blenk et al., "Survey on network virtualization hypervisors for software defined networking," *IEEE Communications Surveys & Tutorials*, vol. 18, no. 1, First Quarter, 2016, pp. 655–685.

9. S. Benington, "From embryonic to tectonic: SDN and the path toward industry transformation," *Lightwave*, May/June 2014, pp. 19–22.

10. B. A. A. Nunes et al., "A survey of software-defined networking: past, present, and future of programmable networks," *IEEE Communications Surveys & Tutorials*, vol. 16, no. 3, Third Quarter, 2014, pp. 1617–1634.

11. R. Jain and S. Paul, "Network virtualization and software defined networking for cloud computing: a survey," *IEEE Communications Magazine*, vol. 51, no. 11, November 2013, pp. 24–31.

12. D. Kreutz et al., "Software-defined networking: a comprehensive survey," *Proceedings of IEEE*, vol. 103, no. 1, January 2015, pp. 14–76.

13. B. Underdahl and G. Kinghorn, *Software Defined Networking for Dummies*. Hoboken, NJ: John Wiley & Sons, 2015.

14. W. Stallings, *Foundations of Modern Networking: SDN, NFV, QoE, IoT, and Cloud*. Indianapolis, IN: Pearson Education, 2016, pp. 76–91.

15. S. T. Ali et al., "A survey of securing networks using software defined networking," *IEEE Transactions on Reliability*, vol. 64, no. 3, September 2015, pp. 1086–1097.

16. M. Chahal et al., "A survey on software-defined networking in vehicular ad hoc networks: challenges, applications and use cases," *Sustainable Cities and Society*, vol. 35, 2017, pp. 830–840.

17. N. Bizanis and F. A. Kuipers, "SDN and virtualization solutions for the Internet of Things: a survey," *IEEE Access*, vol. 4, 2016, pp. 5591–5606.

18. S. Bera, S. Misra, and A. V. Vasilakos, "Software-defined networking for Internet of Things: a survey," *IEEE Internet of Things Journal*, vol. 4, no. 6, December 2017, pp. 1994–2008.

19. A. Thyagaturu et al., "Software defined optical networks (SDONs): a comprehensive survey," *IEEE Communications Surveys & Tutorials*, vol. 18, no. 4, 2016, pp. 2738–2786.

20. P. Göransson, C. Black, and T. Culver, *Software Defined Networks: A Comprehensive Approach*. Waltham, MA: Morgan Kaufmann Publishers, 2017, pp. 191–215.

21. "Software-defined networking," *Wikipedia*, the free encyclopedia https://en.wikipedia.org/wiki/Software-defined_networking.

22. K. Kirkpatrick, "Software-defined networkin," *Communications of the ACM*, vol. 56, no. 9, September 2013, pp. 16–19.

23. M. Dabbagh et al., "Software-defined networking security: pros and cons," *IEEE Communications Magazine*, vol. 53, no. 6, June 2015, pp. 73–79.
24. I. Alsmadi and D. Xu, "Security of software defined networks: a survey," *Computers & Security*, vol. 53, 2015, pp. 79–108.
25. N. A. Jagadeesan and B. Krishnamachari, "Software-defined networking paradigms in wireless networks: a survey," *ACM Computing Surveys*, vol. 47, no. 2, November 2014, pp. 27:1–27:11.
26. A. Akhunzada et al., "Secure and dependable software defined networks," *Journal of Network and Computer Applications*, vol. 61, 2015, pp. 199–221.
27. M. N. O. Sadiku, S. R. Nelatury, and S. M. Musa, "Software defined everything," *Journal of Scientific and Engineering Research*, vol. 4, no. 5, 2017, pp. 48–50.
28. J. Suh et al., "Designing a course and infrastructure for teaching software-defined networking," *Computer Applications in Engineering Education*, vol. 25, 2017, pp. 554–567.
29. N. Dayal et al., "Research trends in security and DDoS in SDN," *Security and Communication Networks*, vol. 9, 2016, pp. 6386–6411.
30. F. Hu, Q. Hao, and K. Bao, "A survey on software-defined network and OpenFlow: from concept to implementation," *IEEE Communications Surveys & Tutorials*, vol. 16, no. 4, 2014, pp. 2181–2206.
31. H. Farhady, H. Lee, and A. Nakao, "Software-defined networking: a survey," *Computer Networks*, vol. 81, 2015, pp. 79–95.
32. Y. Jarraya, T. Madi, and M. Debbabi, "A survey and a layered taxonomy of software-defined networking," *IEEE Communications Surveys & Tutorials*, vol. 16, no. 4, Fourth Quarter, 2014, pp. 1955–1980.
33. W. Xiu et al., "A survey on software-defined networking," *IEEE Communications Surveys & Tutorials*, vol. 17, no. 1, First Quarter, 2015, pp. 27–51.

7

Online Education

Education is the most powerful weapon which you can use to change the world.

Nelson Mandela

7.1 Introduction

Education is widely recognized as a source of human capital which is a useful way of encouraging social and economic development. It is an insurance against poverty. Prior to the digital age, distance learning took the form of correspondence courses, which emphasize individual self-paced lessons. Education providers are moving from traditional face-to-face environments to those that are completely electronic. The movement to online education requires applying new strategies to suit the new medium. As the modernization of distance education and training, online education uses computer networks and satellite communication networks to deliver education. It provides university equivalent courses for millions of students across the globe.

Although some advocates prefer the traditional education delivery methods which include face-to-face communication, there is an alteration going on which affects instruction and learning at the university level. Online education is an emerging field that is situated at the junction of distance education and instructional technology. It is becoming a new paradigm in education.

Online education is an equal opportunity phenomenon, giving the same opportunities to all students, colleges, and universities. It should be regarded as a win–win because it achieves good and opens up new markets simultaneously. A significant number of colleges in the US and abroad are moving from the traditional face-to-face classes into fully online, web-based courses. They are joining the ranks of institutions offering online education.

Online education, often called distance education or web-based education, is currently the latest, most popular form of distance education. It applies the Internet and communication technologies and makes education open, dynamic, and affordable to those who want to learn, regardless of their age or location. More and more universities and publishers world-wide have opted to use online education. For example, for-profit institutions such as University of Phoenix, Kaplan University, National University, Nova Southeastern University, Walden University, and Athabasca University (in Canada) have dominated the online market. Now, there are evening classes, weekend classes, satellite campus, and cyberclasses.

Online courses refer those courses in which at least 80% of the content is delivered online. Massive Online Open Courses (MOOCs) are a subset of online courses. Since the introduction of the first MOOCs in 2003 in the United Kingdom, millions of students across the world have grabbed the opportunity to take courses online. MOOCs help students acquire knowledge in a self-paced manner and choose what they learn and when to learn. They have proved useful for students especially from under-developed countries. They enable them access standard courses for little or no cost [1,2].

This chapter begins with the need for online education. It addresses online teaching, online learning, and online laboratory for courses that require the development of hands-on skills. Then it covers MOOC, which is an open online course that allows anyone to register without time limitations. It discusses enabling technologies for online education, its quality and effectiveness, benefits, and challenges.

7.2 Need for Online Education

The recent development in the Internet coupled with low-cost communication has caused a surge in online education. The use of animation, virtual reality, audio, video, chat, video conferencing, and social networking sites make online learning a rich experience.

Online education is well known for its appeal to nontraditional students. There are some factors driving the need for online education. These include [3]:

- *Maintaining competent workforce.* Due to rapidly changing environment and competition, employee's knowledge and skills need constant updating. Using the traditional methods is expensive. Online education reduces the cost of maintaining a competent workforce.

- *Lifelong learning.* Only few people hold the same job throughout their lifetime. To remain relevant and employable, people need to acquire new skills. Online education provides a cheap, flexible, and convenient means for lifelong learning.

- *Cost.* The cost of traditional education has risen in recent years. Several cost-conscious students choose to take some of their courses online.

- *Convenience.* Online courses are available 24/7 and students can study them in their free, convenient time. They can keep their jobs and study at their own pace. Online education offers flexibility to both instructors and students. Online education is preferred by students who cannot participate in traditional classroom settings.
- *Decline in state funding.* Traditional education is facing a lot of challenges—it is becoming more expensive, there is shortage of professors, cut in funding, busy classrooms, course shortages, limited infrastructures, etc. For more than two decades, state support for higher education has declined. Colleges and universities respond to this by raising tuition, cutting programs, increasing class sizes, and increasing their online course offerings to increase enrolment.

7.3 Online Teaching

Teachers are under pressure to use digital technologies in teaching students and prepare them for work in a globalized digital economy. Online teaching offers exciting opportunities to expand the learning environment for diverse student populations. As the demand for online teaching increases, college professors may be asked to consider teaching their classes online. Online teaching shares much with face-to-face teaching, but it also has a unique set of skills and requirements. Both approaches are similar in content, except in pace and delivery. Rather that developing the courses from scratch, a company has emerged to take care of the courses. Professors just need to use course management system (CMS) software to prepare and deliver their courses. Using the software allows instructors to get it right from the beginning.

For online teaching to be successful, it is recommend that the instructor should follow the following seven principles [4]: (1) encourage student participation, (2) encourage student cooperation, (3) encourage active learning, (4) give prompt feedback, (5) emphasize time on task, (5) communicate high expectations, (7) Respect diverse talents and ways of learning. To these principles one may add seven more [5]: (1) address individual differences, (2) motivate the student, (3) avoid information overload, (4) create a real-life context, (5) encourage social interaction, (6) provide hands-on activities, and (7) encourage student reflection.

Issues facing an online instructor include being effective in delivering the course, responding to student emails, getting used to the online tools and infrastructure. Critics of online teaching and learning question its value, effectiveness, and quality. Since online teaching and learning systems have not been able to convey interactions between the instructor and students; its educational effectiveness is lower than the traditional face-to-face lecture.

Responding to student email messages in a timely manner can be challenging since it requires significant amount of instructor's time. It takes a lot of time to prepare and teach an online course. The challenge of online education largely depends on online instructors. There is also the issue of intellectual property and ownership of materials placed on the web [6].

7.4 Online Learning

Online learning refers to a form of distance learning that takes place partially or entirely over the web. It is learning that is supported by information and communication technologies (ICTs). The "brick and mortar" classroom has started losing its monopoly as the sole place of learning.

The process of learning is complex and it involves the auditory, visual, and tactile senses. The traditional way of learning at a campus university is not for everyone. Online learning is for those who wish to study for a degree alongside work or other commitments. It allows student to pursue an internationally recognized degree without the need to attend classes on campus. It is convenient since it allows one to study anywhere that has an Internet access. They can learn whatever they want, when they want it. The learning is self-paced and offered at a lower cost. Online courses are available 24/7.

Online learning started in 1989 when the University of Phoenix started to offer courses through the Internet. Online learning has been referred to as a form of distance education and as web-based learning, e-learning, and digital learning. It is offered over the Internet and uses web-based materials and activities.

Online courses may be delivered in two ways: asynchronous learning or synchronous learning format. Asynchronous learning takes place when online participants are not required to be online at the same time. It is flexible because it allows instructors and learners to communicate anytime, anywhere. Although students communicate through emails and phone calls, asynchronous learning is regarded as less social in nature and can cause learners to feel isolated. Distance education is customarily asynchronous. Synchronous learning takes place when online class meets at the same specified meeting time and the learners utilize the online media at the same time. It serves four functions: instruction, collaboration, support, and information exchanges. In a synchronous environment, instructors and students can interact with real-time responses. This is close to face-to-face learning [7].

Students need to be technologically savvy to use technology tools that may be required. Students of the digital age appear to be independent, more technology disciplined, and technology savvy, well suited for online environment. Online learning at your own pace is beneficial for a high-quality college degree. It can allow introverts to thrive in ways that traditional setting cannot.

Online learning is under the broader scope of electronic learning (or e-learning). An e-learning process involves technological infrastructure, e-learning software platform, e-learning content, and participants. Popular e-learning platforms include Moodle, Claroline, and EdX. Moodle is a free, online learning management system (LMS) used for blended learning, distance education, and flipped classroom. Claroline is a collaborative online learning and working open-source platform. It is easy to use and available in several countries and languages. EdX is an open-source and free online LMS. Its goal was to act as the WordPress for MOOC [8].

Whether offered on campus or delivered online, each course offering must meet the same rigorous criteria and the strict academic standards. The only difference is in the way the course is delivered. Generally, students are required to have access to a computer system with high-speed Internet connections. They may also expect electronic academic support services such as registration, financial aid, libraries, tutoring, and advisement.

Issues facing online students include the requirement of self-directed learning and self-discipline which may influence the success or failure of online learners. They may be tempted to procrastinate in working on their assignments. The issue of quality in online learning has been raised and it is as complex as the reality of online learning itself. The Quality Matters Program based in the US (www.qmprogram.org) has established national benchmarks for online courses and has become internationally recognized [9].

7.5 Online Laboratory

The online laboratory offers a convenient, less expensive, and more efficient means of providing laboratory experiments to students using the Internet. The experiments can be accessed anytime and anywhere with limited access to a traditional campus setting. Online laboratories are becoming important and common for two reasons. First, laboratories have become indispensable tools for teaching, training, and learning in science, engineering, and technology. Engineering, in particular, is all about hands-on learning of concepts and needs an effective laboratory courseware to complement the theoretical courses. Second, online learning has become a part of the educational landscape. It is growing in scope and acceptance because of the flexibility for the learner and cost-effectiveness for the institution. Online education needs virtual laboratories that can meet the needs of modern science education for their students.

The online laboratory (or laboratory at distance) provides the possibility of students conducting scientific experiments in a virtual environment. Internet-based experimentation permits the use of resources, knowledge, software, and data available on the web. It is bringing laboratory into the home. This can cater to students at the undergraduate and graduate levels. The motivation for developing online laboratory is to make the whole experience of a laboratory more accessible, more convenient, less expensive, and more efficient. Also, providing the resources necessary in the traditional physical laboratory setting is challenged by increasing budgetary and space constraints. Delivery of online laboratory is a potential cost-effective solution to the problem. Compared to their offline (or onsite) equivalents, online experiments are more customizable and scalable.

Online laboratories are no longer just a science fiction dream. Many institutions of learning have started to implement them into their learning process. Online lab is offered in electric circuits, electronics, communication, control systems, computer science, electrical engineering, mechanical engineering, civil engineering, biomedical engineering, physical sciences, medicine, and psychology.

There are two types of online laboratories: (1) Remote lab provides students with the opportunity to collect data from a real physical laboratory setup. It uses real plants and physical devices which are teleoperated in real time. (2) Virtual lab simulates the real equipment. Simulations have evolved into interactive graphical user interfaces where students can manipulate the experiment parameters and do some exploration. The two types of lab can be combined to support specific learning activities.

Motivating Internet users to participate in a web-based experiment can be a resource-intensive and difficult task. Online experiments require a great deal of technical expertise to create and maintain [10,11].

7.6 Massive Open Online Courses

MOOC is an open online course that allows anyone to register without time limitations, geographic restrictions, or prerequisites. MOOCs are essentially a new type of online education that allows anyone, anywhere, to participate via video lectures, peer-to-peer activities, computer graded tests, and discussion forums.

They invite unlimited participation over the Internet. The goal of MOOCs is to provide online education for busy people for the careers of tomorrow and extend knowledge and skills to the entire world. MOOC providers aim at offering the best courses over the Internet, from the best professors and the best schools, covering several disciplines.

MOOC is a form of distance education which was introduced in 2006. The term "Massive Open Online Course" was coined in 2008 by George Siemens and Stephen Downes after carrying out an online course at by the University of Manitoba, Canada [12]. New York Times declared year 2012 as the Year of MOOC because that was when MOOCs hit the mainstream, with private companies including Coursera and Udacity established, and set out to partner with top tier US universities. MOOCs appear to be a collective effort to bring higher education into the digital age. The main characteristics of an MOOC include the following [13]:

- It is massive. There is no limit to the number of learners.
- It is flexible. Learners can participate at various levels.
- It is a web-based education that facilitates learners' empowerment.
- It is easily accessible. Participants have access to additional resources.
- It offers lower costs of education.
- It cuts down the costs of labor and infrastructure.

The major non-profits MOOC providers include Khan Academy and edX, while for-profits MOOC providers include Udacity and Coursera. Other non-academic MOOC providers include Saylor, Udemy, Skillfeed, UoPeople, and Academic Earth. University MOOC pioneers include Massachusetts Institute of Technology, Stanford University, Harvard University, University of Pennsylvania, Georgina Institute of Technology, University of Texas at Austin, University of California at Berkeley, San Jose State University, and Kaplan University [14]. There are several other MOOC providers around the world, in United Kingdom, Canada, India, China, etc. They offer MOOCs in many areas such as engineering, computer science, finance, business, education, health sciences, criminal justice, cyber-security, IT, psychology, archeology, legal studies, and nursing. It is claimed that MOOCs are based on sound pedagogical principles that are comparable with courses offered by colleges and universities in face-to-face mode.

MOOCs have some advantages: democratizing learning, free courses, and economy of scale. The MOOC phenomenon is recent, disruptive, and revolutionary in higher education. Although MOOCs represent a recent, huge step in open education, many issues and questions remain open and need to be addressed. They have come under increased scrutiny over their goals, uses, and effectiveness. Although MOOCs may not displace brick-and-mortar higher education, they are here to stay and provide innovative, lifelong education through professors at prestigious universities in partnership with MOOC providers [15].

7.7 Enabling Technology

Online education needs effective tools to create and deliver content. More than 110 course management software packages have been introduced into the online education market [16]. The work of policymakers, educators, parents, and learners are all being governed by digital technologies. For example, the Internet is providing an easy access to unlimited amount of information. It allows students and teachers to learn different ways of learning. Online courses are best taught when they are engineered to take advantage of the learning opportunities afforded by the digital online technologies. Technology has been a boom for educators and learners. It has been the catalysts for online delivery of higher education. The application of technology and the proliferation of computers, Internet, and smart devices have transformed the way education is delivered and received. Technology is used as a medium for online or virtual education. The main requirement is a computer system and Internet connectivity (cable, wireless, DSL, etc.). Modern technologies such as web-based applications, multimedia, video records, and search engines are used in online education. Using the audio/video-based delivery is the closest way to the traditional face-to-face learning environments. Use of Facebook, Twitter, and Skype chat enhances student-to-student or student-to-professor interactions.

The minimum technological infrastructure required for online education includes computers, computer networks, online student services, and a CMS [17]. The tools required for developing online education includes email servers, web servers, list servers, bulletin boards, chat rooms, audio/video servers, Internet conference servers, and web channel servers [18].

7.8 Presence

In the face-to-face learning environment, we do not need to think much about being present because we are there physically. Creating a sense of presence is very important for an online education because it can enhance instructor-learner relationship. People are social beings by nature. Our sense of presence is experienced in different ways. Presence involves two things: telepresence and social presence. Telepresence occurs when learners feel that they are present at a remote location. Social presence implies that there are interactions with others online. There is feeling of being there with other learners and the instructor.

One cannot be pragmatic in designing instructor presence. Instructor presence includes the planning that goes into developing your course and what you do when interacting with the students. The sense of instructor presence is created through a combination of videos, photos, and narratives, depending on how comfortable the instructors are with the medium. By the middle of the courses, learners should have identified others they feel kinship. It feels that the instructors and learners are accessible to each other.

As technology evolves, we are no longer limited by physical interactions. We are interconnected and our world becomes smaller [19].

7.9 Quality and Effectiveness

Little is known about the effectiveness of online courses in engaging students in the learning process. We do not know whether the course offered is of the highest standard of education. Although most online courses such as MOOCs are well-packaged, their instructional quality is relatively poor. Both administrators and faculty have expressed concern regarding the quality and effectiveness of online education. Even prospective employees are skeptical and concerned about the quality of online educations and the graduates it produces.

Since the primary goal of education is learning, learning effectiveness must be the first measure by which online education is assessed. Online education is not equivalent to traditional education as student learning may not be on same quality. Instructional quality refers to concerns about the effectiveness of teaching or learning environments in the light of educational standards. Advocates of online education need to demonstrate that online teaching and learning are at least as good as classroom education. While some observe worse performance in online courses versus traditional courses, others find the opposite. Thus quality depends on the beholder and the situation at hand. The traditional indicators for quality are changing. As shown in Figure 7.1, the five pillars of quality online education have been identified as: learning effectiveness, student satisfaction, faculty satisfaction, scale (cost-effectiveness), and access [20].

Some institutions use faculty effectiveness as a measure of quality. Technology cannot replace the role of the well-prepared teacher. Others see the student as ultimately influencing an online course effectiveness and performance. Effective learning stems from active participation and involvement of the learner. Healthy participation is energetic, spontaneous, and helpful in motivating students. There is high correlation between instructor-student and student-student interactions in online course and student satisfaction.

7.10 Benefits

The impact of online education is obvious in many fields such as lifelong learning, professional development, and business training. There are several reasons online education is growing in popularity and has attracted

FIGURE 7.1
Five pillars of quality online education [20].

people from diverse groups. Perhaps the greatest, initial appeal for online education is its convenience, accessibility, and availability to learners. Online education improves access to higher education and makes it possible for more people to attend college. It makes the best quality education and best educators available to the whole world. Access to the world's top professors is priceless.

Online education can solve some of these issues. It is scalable and less expensive [21]. It allows students to work on the course anywhere there is Internet connection. There is no discrimination among students on the basis of race, sex, religion, and nationality. A student can exactly study what he or she wants. Since the courses are self-paced, the student can complete it whenever they can.

It eliminates a "one-size-fits-all" approach (which is ineffective) and can be customized to meet diverse learning needs. There is flexible class time and the ability to attend class anywhere there is Internet connection. Because they are unscheduled, online courses increase student autonomy. Students can work full time and earn their degree. Time and money are saved by not driving to school. Courses that are not offered in many institutions can be offered online. Online education has been employed as a means of achieving a balance between the competing demands of family, work, and school.

Online education offers numerous opportunities for both educators and learners. It opens classrooms to the world. Cultures may be shared through online learning and gender is not an issue. Course contents may be translated into several languages. Although online education may not work for everyone, some less-developed countries see the online education as cost-effective.

7.11 Challenges

While online education solves some problems facing traditional education, it poses a different set of challenges. In spite of the proliferation of online courses, some observers have expressed some concerns. First, faculty support is mixed. While some faculty members embrace online education, some resist the shift to online course delivery. They resist supporting and actively participating in online education. They see online education as a potential threat to current models of education and a low-quality substitute for traditional way of learning. It takes more time for a professor to prepare and teach online courses than the regular, traditional face-to-face courses. Long after he goes to bed, students send emails and post messages.

Second, the majority of faculty members are apprehensive of new technologies. They express apprehension about online education because the associated technologies can be frustrating to use. They are not motivated to attend educational technology conferences or read books dealing with creative applications of digital technologies.

Third, assessment, the activity of measuring learning, is a major problem in online education. Course developers often use the simple multiple-choice items in their exams. This is quite inadequate for future leaders, managers, engineers, and scientists.

Fourth, there is the problem with accreditation. There has been a proliferation of colleges and universities offering online programs. Some of these are unwilling to obtain accreditation for their programs. Accreditation agencies never deal with the issues of quality. These are challenges that online educators must not ignore [22].

Fifth, a major disadvantage of online education is the absence of face-to-face or interpersonal interactions with instructors and fellow students. Most students are working full time and have many distractions. There is the problem of assuring the identity of online students.

Sixth, online education has a number of direct and indirect costs, including hardware, software, website development and maintenance, and video recording. Developing a course for online education (for edX and Coursera, for example) may take a professor hundreds of hours and more time in revising it. Students taking online courses complain that it is time-consuming navigating the web.

Seventh, some critics of online education complain that one cannot ensure rigor of the offerings and the quality of the education. Many stakeholders are apprehensive over the lack of quality or richness of online courses. Some for-profit online education providers are more concerned about revenue and enrollment numbers than the quality of their programs. It is harder to verify that students are not cheating in an online test-taking environment. It lacks the moral and ethnical engagement necessary for literal education. Instructors can use tools such as Turnitin to detect plagiarism.

Eighth, the focus of online education is geared toward delivery the information, not the student engagement. What is lacking is the ability to change from data to information, from information to knowledge, and from knowledge to wisdom within each learner [23].

Incorporation of hands-on laboratory experiments in an online engineering course is a challenge. Also, there is the perception that online courses take more time and work; only self-disciplined and self-motivated students are likely to succeed. It is perceived as an extension of the university of capitalism which is now digital and global.

Finally, there are also a set of legal and ethical issues which must be resolved. The legal problem particularly involves copyright and intellectual property. Students' self-disclosure online remains permanent and runs the risks of electronic breaches or hacking.

In developing nations, barriers to online education include prohibitive cost of Internet connections and lack of adequate technical infrastructures.

7.12 Conclusion

Computer networks such as the Internet and digital technologies have transformed education at all levels to meet the demand of the 21st century. Online K-12 schools are spreading across the United States. Schooling that combines computerized learning seems to be the emerging model [24]. With the advances in Internet technologies, online education has gained a lot of popularity in recent times and plays a formidable role in U.S. higher education today. It will continue to become an increasingly mainstream mode of instruction.

Online education has a great potential to reach students with personalized education at a low cost. It has also has the potential of revolutionizing global education and narrowing the gap between developing and developed nations. It is gaining ground as an extension of traditional education. The emergence of social networking technologies is rapidly changing the delivery of online education. By taking advantage of these technologies, online education can provide quality education anywhere, anytime. Online education is growing and traditional colleges and universities should begin to leverage on it. Businesses and corporations are quick in accepting online training. Online education is applied in all disciplines such as engineering, computer science, medicine, nursing, business, music, and social sciences.

Online education is relatively new. It is here to stay and grow. It is exploding in recent years as an option in colleges and universities both within the US and abroad. Most universities and colleges agree that online education is critical to their long-term strategy. It is the future of education. In transforming higher education, it will leave no stone of the institution untouched. It can be predicted that very soon the majority of college courses will use some form of online communications [25].

As the demand for online education by those who have jobs and require lifelong education increases, there are more and more expectations on the implementation of on teaching and learning system [26]. More information about online education can be found be found in books [27–33] and journals exclusively devoted to it: *Distance Education, Journal of Distance Education, Journal of Interactive Online Learning,* and *Journal of Educators Online.*

References

1. I. Literat, "Implications of massive open online courses for higher education: mitigating or reifying educational inequalities?" *Higher Education Research & Development,* vol. 34, no. 6, 2015, pp. 1164–1177.
2. S. K. Ch and S. Popuri, "Impact of online education: a study of online learning platforms and edX," *Proceedings of IEEE International Conference on MOOC, Innovation and Technology in Education,* 2013, pp. 366–370.
3. A. H. Huang, "A supply-chain management perspective of online education," *Journal of Educational Technology Systems,* vol. 29, no. 2, 2000, pp. 93–106.
4. J. Stern, "Introduction to online teaching and learning," www.wlac.edu/online/ documents/otl.pdf.
5. H. Zsohar and J. A. Smith, "Transition from the classroom to the web: successful strategies of teaching online," http://northeast.edu/CTC/Pdf/Successful-strategies-for-teaching-online.pdf.
6. S. Suryanarayanan and E. Kyriakides, "An online portal for collaborative learning and teaching for power engineering education," *IEEE Transactions on Power Systems,* vol. 19, no.1, February 2004, pp. 73–80.
7. "Online learning in higher education," *Wikipedia,* the free encyclopedia https://en.wikipedia.org/wiki/Online_learning_in_higher_education.
8. N. Harrati, I. Bouchrika, and Z. Mahfouf, "e-Learning: on the uptake of modern technologies for online education," *Proceedings of the 6th International Conference on Information Communication and Management,* 2016, pp. 162–166.
9. N. Butcher and M. Wilson-Strydom, *A Guide to Quality in Online Learning.* Dallas, TX: Academic Partnerships, 2013.
10. M. N. O. Sadiku, M. Tembely, and S. M. Musa, "Online laboratory," *International Journal of Engineering Research,* vol. 6, no. 9, September 2017, pp. 425–426.

11. A. Maiti and B. Tripathy, "Remote laboratories: design of experiments and their web implementation," *Educational Technology & Society,* vol. 16, no. 3, 2013, pp. 220–233.
12. T. R. Liyanagunawardena, "Massive open online courses," *Humanities,* vol. 4, no. 1, 2015, pp. 35–41.
13. W. Rubens, M. Kalz, and R. Koper, "Improving the learning design of massive open online courses," *The Turkish Online Journal of Educational Technology,* vol. 13, no. 4, October 2014, pp. 71–80.
14. "Massive open online course," *Wikipedia,* the free encyclopedia https:// en.wikipedia.org/wiki/Massive_open_online_course.
15. M. N. O. Sadiku, S. M. Musa, and S. R. Nelatury, "Massive open online courses," *International Journal of Engineering Research and Allied Sciences,* vol. 2, no. 5, May 2017, pp. 1–3.
16. Y. Kim, "Online education tools," *Public Performance & Management Review,* vol. 28, no. 2, 2004, pp. 275–280.
17. R. L. G. Mitchell, "Online education and organizational change," *Community College Review,* vol. 37, no. 1, July 2009, pp. 81–101.
18. A. Bucur, "Components of online education in gerontology," *Gerontology & Geriatrics Education,* vol. 20, no. 4, 2000, pp. 31–45.
19. R. M. Lehmand and S. C. O. Conceicao, *Creating a Sense of Presence in Online Teaching: How to "Be There" for Distance Learners.* San Francisco, CA: John Wiley & Sons, 2010.
20. K. Shelton, "A quality scorecard for the administration of online education programs: a delphi study," *Doctoral Dissertation,* University of Nebraska, September 2010.
21. J. Harish, "Online education: a revolution in the making," *CADMUS,* vol. 2, no. 1, October 2013, pp. 26–38.
22. T. C. Reeves, "Storms clouds on digital education horizon," *Journal of Computing in Higher Education,* vol. 15, no. 1, Fall 2003, pp. 3–26.
23. S. M. Natale and A. F. Libertella, "Online education: values dilemma in business and the search for emphatic engagement," *Journal of Business Ethics,* vol. 138, 2016, pp. 175–184.
24. M. N. O. Sadiku, A. E. Shadare, and S. M. Musa, "Digital education," *International Journal of Advanced Engineering, Management and Science,* vol. 3, no.1, January 2017, pp. 64–65.
25. M. N. O. Sadiku, M. Tembely, and S. M. Musa, "Online education," *Journal of Multidisciplinary Engineering Science and Technology,* vol. 4, no. 1, January 2017, pp. 6479–8481.
26. M. N. O. Sadiku, P. O. Adebo, and S. M. Musa, "Online teaching and learning," *International Journal of Advanced Research in Computer Science and Software Engineering,* vol. 8, no. 2, February 2018, pp. 73–75.
27. K. C. Cook and K. Grant-Davie (eds.), *Online Education: Global Questions, Local Answers.* Amityville, NY: Baywood Publishing Company, 2005.
28. K. Grant-Davie and K. C. Cook (eds.), *Online Education 2.0: Evolving, Adapting, and Reinventing Online Technical Education.* Amityville, NY: Baywood Publishing Company, 2013.
29. M. A. Maddix, J. R. Estep, and M. E. Lowe (eds.), *Best Practices of Online Education.* Charlotte, NV: Information Age Publishing, 2012.

30. C. Haythornthwaite and M. M. Kazmer (eds.), *Learning, Culture and Community in Online Education: Research and Practice*. New York: Peter Lang Publishing, 2004.

31. J. V. Boettcher and R. M. Conrad, *The Online Teaching Survival Guide*, 2nd edn. San Francisco, CA: John Wiley & Sons, 2016.

32. T. Stavredes, *Effective Online Teaching*. San Francisco, CA: John Wiley & Sons, 2011.

33. M. G. Moore and G. Kearsley, *Distance Education: A Systems View of Online Learning*, 3rd edn. Belmont, CA: Wadsworth, 2012.

Index